Environmental Change, Livelihood Issues and Migration

Avijit Mistri • Bhaswati Das

Environmental Change, Livelihood Issues and Migration

Sundarban Biosphere Reserve, India

Springer

Avijit Mistri
Department of Geography
Nistarini College
Purulia, West Bengal, India

Bhaswati Das
Centre for the Study of Regional Development,
School of Social Sciences
Jawaharlal Nehru University
New Delhi, Delhi, India

ISBN 978-981-13-8734-0 ISBN 978-981-13-8735-7 (eBook)
https://doi.org/10.1007/978-981-13-8735-7

This Springer imprint is published by the registered company Springer Nature Singapore Pte Ltd.
The registered company address is: 152 Beach Road, #21-01/04 Gateway East, Singapore 189721, Singapore

Dedicated
to
the fisherfolk

Foreword

With the climate change, climate-induced migration is a growing concern, especially in the low-lying coastal areas and islands in the developing countries. India is one of the most vulnerable countries to climate change due to high exposure on multiple fronts along with low adaptive capacity. The Himalayas with its vast glaciers in the north and north-east, around 70 million hectares of forest areas with rich natural endowments, and 7,000-km-long coastline with 14.2% (171 million) of Indian population and 13 metro cities expose India to be more vulnerable than any other places. This book deals with empirical data of Indian Sundarban, which is a coastal location, an island setup, and contains magnificent biodiversity which demands out-standing universal values in the front of climate change. Drawing on the field investigation, this book – *Environmental Change, Livelihood Issues and Migration: Sundarban Biosphere Reserve, India* – argues that environmental change in Sundarban affects the livelihood of the islanders and their perceived risk; lack of job opportunities, development deficits, livelihood conflict with environmental legislation, and strategy of migration are widely discussed with empirical rigour. This book is organised into nine chapters and seeks to contribute to the study of climate change and migration nexus which is very challenging due to data limitation and methodology abstruseness. The study reflects the development potential through cyclic migration – a policy intervention to facilitate it on one hand and conservation of biosphere reserve on the other hand. In general, it is a good research work blending both the quantitative and qualitative approach.

I hope that this publication will add to the knowledge of potential scholars, researchers and policymakers regarding the emerging sub-branch of migration study that is environmental migration. I congratulate both the authors, Dr. Mistri and Dr. Das, for carrying on the mammoth task and hope many more works in the future.

Centre for Development Studies
Thiruvananthapuram, Kerala, India

S. Irudaya Rajan

Contents

About the Authors

Avijit Mistri is currently working as Assistant Professor in the Department of Geography, Nistarini College, West Bengal. He completed his postgraduation (M. Sc.) in Geography from Banaras Hindu University, Varanasi, and his M.Phil. and Ph.D. in Population Studies from Jawaharlal Nehru University, New Delhi. He has an extensive work on Sundarban Biosphere Reserve in India in the context of climate change, sustainable livelihood and migration and has published several research articles and chapters in refereed national and international journals and books on this issue.

Bhaswati Das is currently Faculty of Population Studies at the Centre for the Study of Regional Development, School of Social Science, Jawaharlal Nehru University, New Delhi. She completed her graduation in Geography from the erstwhile Presidency College, Kolkata, and continued her higher education in Jawaharlal Nehru University, New Delhi. Her area of research is population and development. Migration being one of the most complex human responses of development, she has published several research papers in peer-reviewed national and international journals on this issue. Several of her research scholars have successfully completed their doctoral thesis addressing issues on migration.

List of Figures

List of Maps

List of Plates

List of Tables

List of Feature Box

Environmental Change and Migration: A Growing Concern

1

Abstract

Environmental migration is being a growing concern all over the world. A gross estimation of 200 million people will be the climate migrants globally by 2050. Low-lying coastal areas and islands in the developing countries are the most vulnerable condition due to low adaptive capacity along with high exposure. The present study area, Indian Sundarban, is a part of the Ganga-Brahmaputra megadelta. Due to the coastal location, an island set up and situated in the tropical cyclonic belt, it is highly exposed to the hydro-climatic hazards. Migration from the Sundarban is a prominent livelihood strategy in almost every household. Inhabitants depend on natural resource-based activities, such as farming and fishing, which are highly susceptible to the climatic stimuli. Moreover, these activities are suffered by seasonality, irrigational and other infrastructural deficiency and conservational rigour. The present study, therefore, is an endeavour to provide insight into the nexus between livelihood issues and migration from the Sundarban where the environment is given special consideration. It is a comparative study based on field investigation of 400 samples which is equally distributed between an exposed (migrants) and controlled (nonmigrants) group.

Keywords

Climate change · Livelihood issues · Environmental migration and adaptation

Around 200 million people will be the climate migrants globally by 2050 (Myers 2005). It is the most accepted figure starting from *Stern Review* (2006) to the UNDP Report (2007). This daunting figure will be overtaken by disruption of monsoon system and other rainfall regimes, by unprecedented severity and long duration of droughts and by sea level rise and flooding in coastal areas (Myers 2005). The utilisation of the coastal region has been increased dramatically during the twentieth century, and it seems certain to keep it up through the twenty-first century.

A. Mistri, B. Das, *Environmental Change, Livelihood Issues and Migration*,
https://doi.org/10.1007/978-981-13-8735-7_1

Approximately 23% of the worlds' population is estimated to live both within 100 km. distance of the coast and less than 100 m above sea level, and the density is three times higher than the global average (Small and Nicholls 2003). Sixty percent of the world's 39 metropolises with a population of above five million are situated within 100 km. of the coast which includes 12 megacities (population above 10 million) out of 16 in the world (IPCC 2007). Nearly 300 million people with an average density 500 people per sq. km. inhibit in 40 deltas globally, including all the megadeltas (Ericson et al. 2006). Among these, the most populated delta is the Ganges-Brahmaputra Delta, and highest density is in the Nile Delta. Using the coarse digital terrain model and global population distribution data, Ericson et al. (2006) estimated that more than one million people will be directly affected by sea level rise in 2050 in each of the three megadeltas, namely, the Ganges-Brahmaputra-Meghna Delta in Bangladesh and West Bengal, the Mekong Delta in Vietnam and the Nile Delta in Egypt. In the estimation of different emission scenarios, IPCC (2000) predicted that by the end of the twenty-first century, the sea level will be 40 cm higher than today resulting in flood victims in coastal areas to increase to 94 million from 13 million per annum. Almost 60% of this uptrend will be confined along the coast from Pakistan through India, Sri Lanka and Bangladesh to Burma (Wassmann et al. 2004). A large population in Vietnam, Bangladesh, parts of China and India is living in low-lying coastal areas and islands. One-quarter of Bangladesh's population (around 35 million) lives in the coastal floodplain (Stern 2006). China's 41% population, 60% wealth and 70% of its megacities are located in coastal areas (UNDP 2007). One-metre sea level rise will inundate 5,763 sq. km. area in India (TERI 1996) and displace approximately 7.1 million people (Bhattacharya 2007). Likewise, a 30 cm sea level rise will inundate 81,348 sq. km. of coastal lowland in China (Du and Zhang 2000). Moreover, marine and coastal ecosystems in Asia are likely to be affected by sea level rise, and temperature increases at high confidence (IPCC 2007). The loss of coral reefs in Asia is projected as high as 80% where the global average is 59% in the next 30 years under IS92a emission scenario (IPCC 1992). A substantial proportion of mangroves that is approximately 45% of the world's mangrove forests colonise in tropical and sub-tropical coasts of Asia. It is estimated that nearly 33% coastal wetlands, including salt marshes and mangroves, will be lost globally by 36 cm rise in sea level during 2000–2080 (IPCC 2007). Study (Mitra et al. 2010) on famous mangrove block in Indian Sundarban revealed that excessive salinity stunted the growth and some mangrove species have gradually been losing, resulting in loses of habitats of the world famous Royal Bengal Tigers. With a sea level rise of 28 cm in Bangladesh, Sundarban declines the habitat of the Royal Bengal Tiger by 96% (Loucks et al. 2010). Apart from the declining forest production, climate change severely affects the production of fish which is the sustenance of 85% fishermen and fish farmers in Asia. In addition to that, coastal areas are severely ravaged by extreme climatic events such as cyclone and storm surge. Every year on an average, 120 million people were affected by tropical cyclone hazards that killed 250,000 people during 1980–2000 of which 60% were from Bangladesh (UNDP 2004). Another study estimated that Bangladesh and India share 49% and 22% of the world's total deaths, respectively, due to cyclonic storm (Khan et al. 2000). Both the frequency and intensity of tropical cyclones in the Pacific Ocean have increased over the last few decades (Fan and Li 2005). Likewise, both the frequency and intensification rate of severe cyclonic storm (SCS) in the Northern Indian Ocean (Bay of Bengal) have registered significant increasing trend in past 122 years during the intense cyclone months, May, October and November (Singh 2007; Singh et al. 2001; Lal 2001; Khan et al. 2000).

The present study area, Indian Sundarban, is a part of the Ganges-Brahmaputra Delta comprised of 102 islands. It is the largest halophytic mangrove forest in the world. It contains magnificent biodiversity and outstanding universal value and is famous for in-situ conservation of globally endangered species. It is a part of the World Network of Biosphere Reserves under UNESCO's Man and the Biosphere Programme (in 2001). A total of 9600 sq. km., including 4200 sq. km. forested area and 5400 sq. km. inhabited area, is recognised as Sundarban Biosphere Reserve. Around 4.4 million people reside in the biosphere reserve. They primarily depend on natural resource-based activities like farming (59%) and fishing (33%). Geographically, the Sundarban is an archipelago (54 inhabited and 48 forested islands) along the Bay of Bengal coast. Most of the inhabited islands are still in juvenile stage. Before completion of the build-up processes through siltation, people erected mud-built embankment and occupied the land from river grasp and settled down. The absolute height of the islands varies from 0.90 to 2.11 metres above the mean sea level (MSL) (DPD 2009). The islands, therefore, are highly susceptible to the global warming as well as hydro-climatic hazards. Nearly 7% of the tropical cyclone of the world is formed over the Bay of Bengal and Arabian Sea in the Northern Indian Ocean, where the ratio of their frequency is 4:1 (Khan et al. 2000). This region has witnessed the recent occurrence of severe cyclonic storm (SCS), Aila in 2009 and Sidr in 2007 (super cyclonic storm), Bijli in 2009 (SCS) and Phailin in 2013 (very SCS). At the fag end of 2000s, IPCC (2007) in its 4th Assessment Report by Group-II warned that Asian megadeltas especially the Ganges-Brahmaputra Delta in Bangladesh and West Bengal in India are in the most vulnerable condition due to climate change, sea level rise and more intensification of coastal storms. This vulnerability will be increased with low human adaptive capacity and high level of exposure. The Sundarban people depend on farming and fishing activities which are highly susceptible to the climatic stimuli. Upsurge of one cyclonic storm becomes the causes of misery minimum 2–3 years complete crop failure and subsequent two more years of low yielding rate due to the high level of salinity; meanwhile, it demolishes the infrastructure of the inland fisheries. Climate change and livelihood issues along with an acute development deficit in the Sundarban (Kanjilal et al. 2013; Kanjilal and Sarkar 2009; Ghosh 2012; Jalais 2011; Banerjee 1998) make the livelihoods more complex and vulnerable. On the other hand, out-migration from the Sundarban is the most common livelihood strategy in each and every household. Data limitation is one of the major constrains of the migration study in the Sundarban as the biosphere reserve is partially included into two administrative districts, namely, South 24 Parganas (S 24 Pgs) and North 24 Parganas (N 24 Pgs), of the Indian State, West Bengal. Hence, intensive study on migration is very spare. However, the study carried by the Jayaprakash Institute of Social Change (JISC),

Kolkata, with reference period from September 2010 to January 2011, revealed that at least one member from 75% respondent's family (out of 500 sample) migrated in search of work (Ghosh 2012). Nearly 73% respondent opined that lack of job opportunities was the key driver for migration from the Sundarban. Around 72% respondent expressed that there was a huge wage rate differential between the place of origin and destination. In addition to that, nearly 80% migrants sent remittances to their family regularly. The studies by Mistri (2012, 2013) capturing migration during almost the same reference period unveiled that around 74% household (out of 200) reported migration; on an average, one person migrated from each household. Studies also revealed that search for job was the key reason for migration which was shared by 86% migrants. Migrants highly preferred interstate migration over interdistrict (intrastate) and intra-district migration. Around 64% people migrated outside their home state (West Bengal), whereas interdistrict and intra-district migrants share 32% and 3%, respectively. Both the studies, JISC and Mistri, revealed that after Aila in 2009, the rate of out-migration had dramatically increased due to the lack of job opportunities at the local level. The present study is an attempt to provide insight into the nexus between livelihood issues and migration from the Sundarban where the environment is given special consideration.

1.1 Environmental Change, Natural Resource-Based Occupations and Migration

Around 70% people in India directly depend on land-based occupations, forests, wetlands and marine habitats (Kothari and Patel 2006). People those who are engaged in these activities are more vulnerable to the climatic vagaries compared to those who are engaged in out of these or nonfarm sectors (Banerjee et al. 2011; McDowell and de Haan 1997). Off-farm earning reduces the uncertainty of family income to the climatic vagaries and helps to invest for the adaptive measures resulting in deterring the propensity of migration (Tiffen et al. 1994). However, studies suggest that the impact of climate change on natural resource-based activities influences the decision of migration differently. It is a century back traditional practice that searching for new productive land clearly promotes migration. In an agrarian economy, switching from the farming into other economic sectors pertains to the geographic mobility (Todaro 1969). An area with less amount of farmland leads higher rate of out-migration than where the land is more abandoned (Shivakoti et al. 1999). A study on Nepal's Chitwan Valley in the context of climate change and migration (Massey et al. 2007) reveals that declining agricultural productivity, falling share of neighbour covers in flora and increasing the time required for collecting the firewood induce the members of the households to leave the home searching for opportunities outside of the valley. Husbandry of large and small animal is a part of food production and considered as stock resources of households in the rural areas of the developing countries. But due to the extremely high rate of deforestation and declining grazing land since 1990 (FAO 2016), access to fodder for animals becomes a great concern. More time and labours are engaged to gather the fodder that motivates the members to relocate to the areas closer to the forests and vegetation. Hence, those families face less access to fodder lead higher rates of out-migration (Massey et al. 2007).

On the contrary, a set of studies shows the positive association between declining farming activities due to environmental change and propensity of migration from that area. A study by the International Centre for Integrated Mountain Development (ICIMOD) in 2011 on water hazard on the Hindu Kush Himalaya observed that the members of households whose agricultural land had been damaged were less likely to migrate compared to the households whose land had not been affected (Banerjee et al. 2011). The reasons were examined that the households whose land were affected by water hazards had tried to repair or mediate the damage during or just after the hazard and they received the governmental compensation as an alternative source of income; absentees were excluded from benefits. The study also investigated that the farming- and animal husbandry-dependent households were also less likely to send their member to migrate out at the time of drought. In a drought condition, additional labourers are required to collect the water for animal husbandry or to irrigate the fields. Moreover, the pattern of migration is differently influenced by the environmental stressors along with the deterrent of migration. A study (Findley 1994) on Mali during the chronic drought condition from 1983 to 1985 revealed that the average rate of migration did not increase and the pattern of migration shifted towards more short-cycle or circular migration (nearly 64%) from the classic long-cycle or permanent migration pattern. Only woman and child migration dramatically increased during that time, which contributed the most to the circular migration. Likewise, it was observed that, in 1972 Nicaraguan earthquake and in 1976 Guatemala earthquake, the migration rate among disaster-affected communities was similar to overall migration rates and the majority of victims not only returned to their homes but the retention rate was also as high as 90% (Perch-Nielsen 2004; Belcher and Bates 1983). Nearly 144 million people were displaced by sudden-onset disasters in the last 5 years globally (Ferries 2014). The sudden-onset disasters like wind storm and hurricanes are likely to lead to temporary distress migration first, and after that, people typically return to the disaster site to rebuild their houses, repair infrastructures and make over livelihood (Raleigh et al. 2008).

Till date, forest resources are the base of survival and livelihoods of a large number of people in India. The people, especially communities, who depends on the natural environment at their locality to meet most of their material needs is termed as 'ecosystem people' (Kothari and Patel 2006). Over the years, environmental legislations which have been adopted either on the ground of environmental change or its related issues in different part of India have immensely affected the livelihood of ecosystem people by forcing them to move out from there. The examples could be drawn as the evidences like the eviction of the nomadic pastoralists, Gujjars, in the Rajaji National Park in the Himalayan foothills by the Forest Department of Uttarakhand (Kothari and Patel 2006). Another example is that the total ban on overgrazing of livestock of the local communities in the wetland Keoladeo Ghana National Park in Bharatpur, Rajasthan (Kothari and Patel 2006). The suspension of Goa mines by the Ministry of Environment and Forests (MoEF), Government of

India (GoI), in 2012 and later revoking the suspension after 3 years in 2015 have hampered the livelihoods of thousands of mining labourers (Aggarwal 2015). Agitation and political unrest are still continuing over the Western Ghats in Kerala since the issuing of official memorandum based on the Gadgil and Kasturirangan Committees' reports by the MoEF (2011, 2013) in 2013. In 1999 and 2000, the conflict turned into retaliation between the local fisherfolks and commercial farm owners in Chilika Lake in Odisha is a classic example of the conflict between the commercial sector and ecosystem people (Kothari and Patel 2006). The eviction of residential villages from the Sariska Tiger Reserve in Alwar, Rajasthan, in 2005 is the most common examples of livelihood conflict in the context of wildlife conservation in India. Likewise, there is a serious alligation observed to the utilisation of forest by the islanders in the present study area, the Sundarban. Nearly 0.5 million people in the Sundarban depends on forest, either fishing or non-timbre forest production (NTFP). The present study provides a detailed illustration of the livelihood conflict with environmental conservations.

1.2 Migration and Sustainable Livelihood

A livelihood comprises the capabilities, assets (stores, resources, claim and access) and activities required for means of living (Camber and Conway 1992). People's basic needs are food, cloth and shelter. 'Activities' are what people do to meet those needs. It encompasses a wide range of occupations in different sectors. A livelihood will be sustainable when it can cope with and recover from stresses and shocks; maintain or enhance capabilities and activities; and provide livelihood opportunities for the next generation (Serrat 2008; Scoones 1998; Camber and Conway 1992). Sustainability of a livelihood is a function where capabilities, assets and activities are utilised, maintained and enhanced in such a way that preserves the livelihood. A livelihood with sustainable character is inevitable for the decent living of the human being. As migration is one of the important livelihood strategies or outcomes (DFID 2001), sustainable livelihood becomes an integral part of the migration study. To understand the linkage between migration and sustainable livelihood, it is not only important to observe the migration in the viewpoint of conventional migration theories, such as poverty, intersectoral income differentials and cost and return (de Han 1997; Todaro 1969; Sjaastad 1962), but also from the viewpoint of inequality, satisfaction or deprivation, frustrations or elations and aspiration are important ones. Studies on Indian villages (Lipton 1980; Connell et al. 1976) during the 1970s observed that unequal, not the poorest, villages had recorded the highest rates of out-migration. Likewise, from the observation of large numbers of villages in less developed countries, Stark (1984) observed three interesting facts; these are (i) rural-urban migration rates are not highest from the poorest villages; (ii) migration rates are higher from villages where the distribution of income by size is more unequal; and (iii) from these villages, propensity to migrate is the highest for those who are very poor. People are engaged to compare their income from their neighbour or community and internalise it which generates psychic costs or benefits, frustrations or elations, relative deprivation or satisfaction. These factors motive locational decisions whether one person will migrate or stay. Relative deprivation is one of the important factors affecting people's choice of migration (Stark 1984). If a person's (X_1) income (I) is less than the community's average income ($\bar{I}$), it is considered as relatively deprived ($I < \bar{I}$). It is assumed that by migrating to the urban sector, X_1 will secure an income closer to the $\bar{I}$ (Stark 1984). When X_1 migrates, the average income in the village will rise and be greater than $\bar{I}$. Then a villager (X_2) whose income prior to X_1 migration was equal or greater than $\bar{I}$, now, he feels relatively deprived, and he (X_2) also decides to migrate. Again, community's average income rises, and someone else feels deprived and decides to migrate. Thus, this process is continuing over the period. Stark (1984) denotes it as 'self-perpetuating tendency' of migration. In addition to that, the level of deprivation is not only dependent on individual factors, but household context also plays a crucial role. Households' assets, social network and status within a community are integral part of the relative deprivation or satisfaction.

Apart from the relative deprivation, aspiration is also one of the important determinants of migration. Aspiration encompasses a wide range of aspects such as subjective wellbeing, economic wellbeing, aspiration to live in urban areas, aspiration to work in nonfarm sectors, etc. There are four sources of aspiration, viz. first, fortune of being born with a natural capacity to aspire; second, fortune of being born into a prosperous household which can provide necessary economic, social and human resources to develop an aspirational personality; third, being young, well-educated and exposed to socioeconomic development; and fourth, processing proactive behaviours, such as propensity to migrate and switching to the more secure activities, which can further increase aspirations for future (Czaika and Vothknecht 2012). Studies (Czaika and Vothknecht 2012, 2014) based on longitudinal data, Indonesian Family Life Survey (IFLS) in 1997, 2000 and 2007, reveal that migrants have higher aspirations than non-migrants in the case of internal migration. The higher aspirations among the migrants are partly the result of their prior migration experience. In higher levels of aspirations, according to which migrants are self-selected, about 70% are explained by the aspiration-enhancing characteristics like young age, good education and superior socioeconomic backgrounds. A large-scale survey around 500 samples aged 18–39 years in four different areas of Senegal, namely, Darou Mousty, Lambaye, Golf Sud and Orkadiere, unveils that the most of the young adults (64–82%) wish to emigrate from Senegal within the next 5 years (Carling et al. 2013). Likewise, another study (IOM 2011) on the aspirations of Egyptian youth aged 10–29 years based on the sample Survey of Young People in Egypt (SYPE) in 2009 reveals that males are more likely to migrate compared to females and the rate of migration decreases with age. With the increase of wealth and level of educational attainment, the proportion of migration rises. Egyptian youth overwhelmingly prefers to go to Middle East countries in comparison with Western countries, because Middle East countries are easily accessed destinations and cultural similarities persist. In a study on interstate migration in Indian context, Reja (2016) has found the replace migration in the state of Kerala from the West Bengal. Most of the Keralites migrate to the Middle East countries, whereas Bengali migrants fill up the gaps in the workforce. The young and educated migrants among Muslim communities in the district of Murshidabad of the state of West Bengal, India, highly preferred to migrate to Kerala due to the cultural resemblances that facilitate easily access of food, lodging and performance of religious practices.

Another important social dimension of migration is social network. Social network reduces the costs of migration, such as monetary cost, information and search cost, opportunity cost and physic cost, and maximises the benefit as well as reduces the risk at destination. A migrant is more likely to be employed in high-paying jobs in nonagricultural sectors when his network is exogenously larger (Munshi 2003; Montgomery 1991). Network is a very useful capital among the low and unskilled workers, or even, among the illiterate and semi-literate workers. Studies suggest (Iversen et al. 2009; Munshi and Rosenzweig 2006) that there is a high prevalence of referrals in low and unskilled occupations. A study (Munshi and Rosenzweig 2006) based on 4900 households in Mumbai revealed that around 70% of unskilled and blue-collar workers found their first job through a referral, where around 44% white-collar workers used the same. Drawing on the Egypt Labour Market Survey (ELMS 98), Wahba and Zenou (2005) reported that the illiterate and semi-literate workers who were unable to read job advertisements and vacancy posters were strongly incentivised to search works through friends and contacts with intermediaries or contractors. Therefore, there is a negative relationship between network use and the skill intensity of jobs observed in the urban labour markets (Iversen et al. 2009). Referral-based recruitment helps the employer's screening processes and reduces the uncertainty about the candidate's labour productivity meanwhile incentivises the loyalty to the workplace. It is, most often, found that more established members at destination contribute disproportionately to the network which enormously helps the disadvantaged members like women, the elderly and the less educated workers (Munshi 2003). Over the time social network gradually increases and spreads; once, it has reached a critical level when every person of a community close link to someone with migrant experience, then the cost of migration drastically reduces, and place of destinations are facilitated by cheap labour force (Massey 1990). Moreover, in the digital era, just with a click, we can easily get plenty of information about the place of destinations, or we can get thousands of friends at a fraction of second through social network sites. Modern information and communication technology (ICT) make the social network stronger and wider. The mobile phone is one of the most popular devices among them. Nearly seven billion mobile phones are used in the world, whereas India possesses 1.03 billion mobile phones just after China 1.23 billion. Use of mobile phone and likelihood of migration is positively related. Study on the impacts of mobile phones and personal networks of rural-urban migration in Uganda by Muto (2012) reveals that possession of one or more mobile phones or handsets in a household increases the chance of leaving its member to find a job out of villages. The use of mobile phone increased the chances of migration to a greater degree to the workers, those who belong to a smaller ethnic group in comparison to those who belong to a larger group in the capital city, Kampala in Uganda. Finally, it can be concluded that migration is a multicausal phenomenon, not too easy to generalise by the conventional migration theories like poverty, wage differentials and cost and return. The sustainable livelihood, therefore, is one of the best approaches of migration study, which focus on the complexity or multidimensionality with local relevancy. The present study highly accentuates on sustainable livelihood approach where out-migration is explored with multiple dimensions pertaining to environmental, socioeconomic and demographic factors. Meanwhile, the characteristics of migration, such as pattern, stream, seasonality and circularity, are also discussed with empirical rigour.

1.3 Evolution of Environmental Migration

Movement of people as a result of changes in the environment is not a new one. Archaeological evidences suggest that the Egypt and Mesopotamia civilisation emerged as people migrated away from desiccating rangeland and into riverine areas (Brooks 2006). In response to the changes of environment, they were forced to build up the densely packed population in restricted areas in order to manage scarce resources resulting in emergence of the first large urban society. Besides, the nomadic people and pastoralists have been moving seasonally in response to a change of their environment for centuries. However, it is only in the last 20 years or so that the international community has begun to slowly recognise the influence of changing climate on human mobility (Laczko and Aghazarm 2009). At the beginning of the 1990s, the IPCC (1992) warned that the greatest single impact of climate change could be on human migration with millions of people displaced by shoreline erosion, coastal flooding and severe drought. Next, in 1992, International Organisation for Migration (IOM) together with the Refugee Policy Group published a report on 'Migration and Environment' in which it was stated that environmental migration would increase substantially owing to become uninhabitable of the wide geographical area in the earth in the coming years. IPCC developed long-term emission scenarios in 1990 and 1992 and latest in 2000. These scenarios had been widely used to analyse the possible change of climate, its impacts and measures of mitigation, but human mobility and its wide consequences were not brought out in the limelight. In the middle of the 1990s, there was some effort made to estimate environmental-induced migrants that were nearly 25 million and estimated to be double in between 1995 and 2005 (Myers 2005). That time it was assumed that all environmental effects would result in displacement migration and too often defined as a refugee. At that time the issue was framed within a wider security debate, but the momentum did not last.

After a gap, in the middle of the 2000s, the nexus between environment and human mobility again gain its momentum. From IPCC's assessment reports (2007), UNDP's HDR report (2007) and *Stern Review* (2006) to different studies (Massey et al. 2007; Ericson et al. 2006; McLeman and Smith 2006; Myers 2005; Small and Nicholls 2003; Wassmann et al. 2004) highly emphasise on the great impact of environmental change on human being. Around 200 million environmental migrants by 2050 are projected worldwide. In 2010, first time IOM tried to define the environmental migration, conceptualise its framework and develop a guideline to overcome the data collection and methodological challenges. According to IOM (2009, p. 19) definition 'environmental migrants are persons or groups of persons who, for compelling reasons of sudden or progressive change in the environment that adversely affects their lives or living conditions, are obliged to leave their habitual homes, or choose to do so, either temporarily or permanently, and who move either within their country or abroad'. Still, the study of environmental migration has been facing lots of challenges in terms of wide applications empirically. Part of the reason may be hidden into the discipline itself (Laczko and Aghazarm 2009), viz. first, difficulties to find out the evidences to conceptualise the impact of climate change on migration; second, data collection and methodological challenges and approaches; and third, policy responses and legal frameworks. While the scientific argument for climate change is increasingly confident, the consequence of climate change for human population distribution is unclear and unpredictable (Brown 2007). Any environmental migration study begins with a fundamental question that will migration be influenced by climatic change in a linear manner or a nonlinear manner? There are little evidences of the direct influence of climate change on contemporary migration pattern (ADB 2009). Climate change interacts with the wide range of economic, social and demographic factors which influence the decision of migration. In this context, Shurke (1993) points out that the research literature on environmental migration is divided into two broad categories: first, work done by 'minimalists' who suggest that the environment is only a contextual factor in decision of migration, and second, work done by 'maximalists' who argue that the environment directly forced the people to move out. These evoke the duality between two traditional geographical thoughts, environmental determinism and possibilism. There is wide consensus that it is important to view the environment as a cause of migration along a spectrum (ADB 2009). It is found that people are immobile in an extreme weather event or moving voluntarily, while other people are forced to flee out their usual place of residence in the same environmental stressor (Black et al. 2012). People have a wide range of possible strategies along with taking or not taking the decision of migration in response to the impact of environmental change. The number of possibilities (strategies) depends on how much a household is endowed in terms of social, economic, cultural and political capital.

However, the impact of climate change is not homogenous and varies regionally. The micro level study gives us more detail explanation of that. Data limitation as the evidence of migration as well as climate change is the main challenges for the micro level study. Therefore, most of the study on environmental migration is somehow inclined towards the macro region study where the regional uniqueness is hardly captured. Moreover, studies related to the nexus between environmental stressors and migration through livelihood issues are very sparse till the date. In these contexts, the present study in the Sundarban is an important one. On the other hand, a plethora of studies (Hazra et al. 2010; Mitra et al. 2009; Singh 2007; Unnikrishnan et al. 2006; Singh et al. 2001; Lal 2001) on changes of environmental parameters like sea level rise and sea surface temperature, rainfall, coastal erosion, cyclonic storms, salinity, water pH, transparency and so on is carried on in the Sundarban. But, its impact on the human being and their livelihoods has not been examined empirically. The present study, therefore, is an attempt to get insight into the subjective appraisal (risk perception) of the environmental change in the Sundarban; its influence on means of living, especially on farming and fishing; and finally its linkage to out-migration. It is a comparative study between two groups, migrants (exposed group) and nonmigrants (control group), which examines the likelihood of migration in reference to the non-migration.

1.4 Conceptual Framework of the Study

The present study is conceptualised (Fig 1.1) on the basis of Sustainable Livelihood Approach (SLA). The central point of the concept is the various asset possessions by a household and its members. The livelihood assets are roughly divided into five categories, such as natural, physical, financial, human and social assets, and these are not mutually exclusive. On the other hand, the stressors in the Sundarban can be divided into two categories, such as environmental and non-environmental. Environmental stressors include both climatic events (sudden onset) and climate processes (progressive). Non-environmental stressors are transforming infrastructures and processes. Processes include rules, regulations and policies. Infrastructure comprises various governmental and non-governmental institutions or organisations. The environmental stressors adversely influence to the livelihood capitals. Hence, people perceive environmental risk in their livelihood. Unlike natural science, the impact of the environment on livelihood is hardly possible to explore directly. The method involved in the present task examines the association is the risk perception study. Non-environmental stressors and livelihood assets interact with each other, and it may enhance or deterrent the efficiency of the assets. Non-environmental stressors also influence the environmental threats. Through the

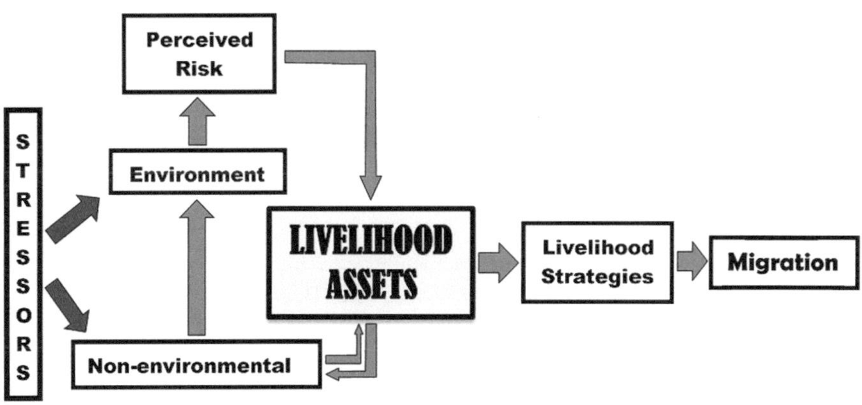

Fig. 1.1 Conceptual framework of the study

interaction between livelihood assets and stressors, different sorts of livelihood strategies or options emerged like job diversification and transformation, migration and displacement, sustainable use of resources and adoption of technologies and reform the policies and institutions and so on. In the Sundarban, the out-migration is the most predominant outcome among them.

The study follows bottom up or inductive approach, which starts from the end point that is strategy of migration. Next, the migration is examined under the purview of livelihood issues, nexus with environmental threats, influence of institutions and policies, and examined other influencing factors. Finally, it ends up with the general conclusions.

1.5 Data Sources

Both the primary and secondary data sources are taken help for the study. Secondary data that included multiple sources ranges from various departments of state and central government, Panchayati Raj Institutions (PRIs), Block Developments Offices (BDOs), Subdivisional Offices (SDOs), forest ranger offices and published and unpublished reports, articles, thesis and dissertations to different non-governmental organisations (NGOs). The major sources of secondary data are as follows in detail:

- Census of India 2001 and 2011, Office of the Registrar General of India (ORGI), Ministry of Home Affairs (MoHA), Government of India (GoI): B-Series (socio-economic data) and D-Series (migration data), Primary Census Abstract (PCA) and Village Directory are used for analysis.
- Agricultural Census of India, Department of Agriculture, Cooperation and Farmers Welfare, Ministry of Agriculture and Farmers Welfare, GoI: Agricultural Census with reference years 2000–2001, 2005–2006, 2010–2011, 2015–2016 and Input Survey with reference years 2001–2002, 2006–2007 and 2011–2012 are accessed from the official website.
- District Statistical Handbook of South and North 24 Parganas, Bureau of Applied Economics and Statistics (B.A.E. & S.), Department of Statistics and Programme Implementation, Government of West Bengal: The B.A.E. & S. accumulates wide range of date sets from socioeconomic infrastructure to climatic data from different governmental sources and publishes a handbook for all the districts of West Bengal. A handbook of a particular district provides data up to Community Development Blocks (C. D. Blocks) level.

Apart from these, other important data sources are the Department of Sundarban Affairs, Government of West Bengal; Sajnekhali Beat Office in the Sajnekhali Wildlife Sanctuary, Sundarban Tiger Reserve; and 18th Livestock Census, Department of Animal Husbandry Dairying and Fisheries, Ministry of Agriculture and Farmers Welfare, GoI.

Primary data are collected through filed investigation. Structured questionnaire schedule are used to collect the data. Total 400 households (HHs) are surveyed. The sample is divided into equal groups such as control group (200 HHs) and exposed group (200 HHs). The control group is that group of households where not a single member had migration experience during the last 365 days before the date of survey. The exposed group is that group where at least one member of the household experienced or was experiencing migration before the date of survey. The survey was conducted in between 5 October and 31 December 2014. All the respondents belong to working age, 15–59 years. In the control group, the head of the household or any working person in the absence of the head of the household was considered as a respondent. In the exposed group, the only migrating person who was present at the time of the survey and bearing the migration status till date, which means he or she came home to an occasion or a purpose but not returned permanently, was considered as a respondent. If more than one presence, the elderly migrant was selected.

1.6 Sample Design

Census villages are considered as a least geographical unit of the sampling, and a respondent in a household within a census village is considered as a least social unit. Total 8 census villages from 4 administrative blocks of South 24 Parganas and a total 50 households, including 25 migrants and 25 nonmigrants, from each census village are surveyed. Total 102 islands comprise the Sundarban Biosphere Reserve. Inhabited islands and forest islands are situated side by side where rivers act as a buffer. Inhabited islands (54 islands) constitute the western part of the Sundarban, and forested islands (48 islands) are located in the eastern part of the Sundarban. A total 19 C. D. Blocks, 13 from the districts South 24 Parganas and 6 from the district North 24 Parganas comprise the inhabited part of the SBR. The Kolkata urban agglomeration highly influences the six blocks under the North 24 Parganas. These are socioeconomically different from the rest of the Sundarban. Therefore, North 24 Parganas is excluded from sampling. Moreover, the maximum portion of the Sundarban that is around 73% of the inhabited area is under South 24 Parganas. It bears typical characteristics of the Sundarban. For the purpose of sampling, the blocks under South 24 Parganas are roughly divided into two physiographic divisions, namely, the northern Sundarban and southern Sundarban (Map 1.1). The blocks, Canning-I and II, Jaynagar-I, Kultali, Basanti and Gosaba, roughly comprised the northern Sundarban, and the rest of the blocks are included in the southern Sundarban. Two blocks from each physiographic division are selected, viz. one adjacent to the forest and another far from the forest and the highest proportion of workers engaged in farming. The C. D. Blocks, namely, Gosaba and Basanti from the northern division and Namkhana and Mathurapur-II from the southern division, are selected. Although agriculture is highly practised in all the blocks of South 24 Parganas, the blocks adjacent to the forest fishing are widely practised compared to the blocks far from the forest (Singh et al. 2010). Hence, the blocks adjacent to the forest, two census villages which are close proximity to the forest, are selected. Likewise, two census villages with the highest

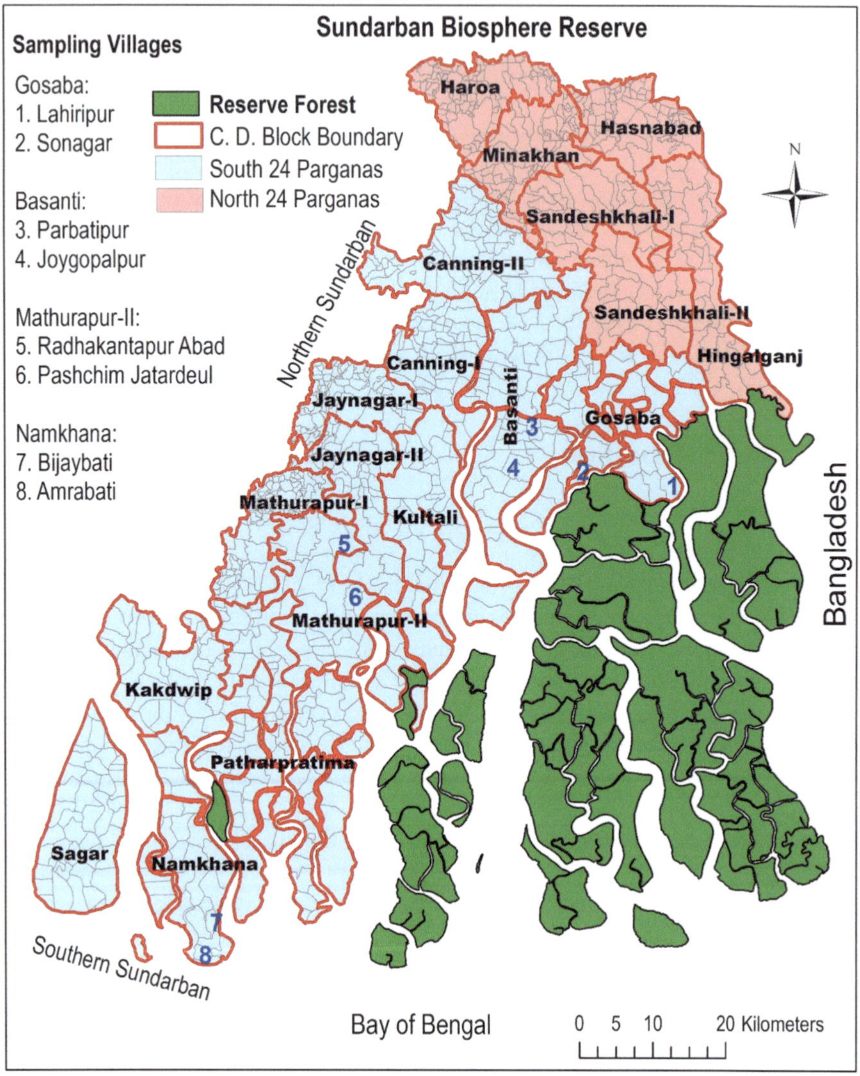

Map 1.1 Map of the sample units

proportion of workers engaged in farming in the (selected) blocks far from the forest are picked up. Thus, total eight census villages from four blocks are designed for sampling. The selected census villages are Lahripur (1) and Sonagar (2) from Gosaba block; Parbatipur (3) and Joygopalpur (4) from Basanti block; Radhakantapur Abad (5) and Pashchim Jatardeul (6) from Mathurapur-II; and Bijaybati (7) and Amaravathi (8) from Namkhana block.

1.7 Methods

Both qualitative and quantitative methods are incorporated in the study. The statistical techniques like chi-square (x^2) test, *t*-test (one sample and independent sample), composite indices with the help of standardisation and principal component analysis (PCA), binary logistic regression and descriptive statistics, such as mean score, mean difference, standard error of mean (SEM) and effect size, are taken help for the data analysis. In addition to that, different migration and demographic measures and cartographic techniques are incorporated wherever these are required.

Some psychological techniques, such as psychophysical scaling with the help of Likert scale and bipolar affect scale, risk perception indices, the area of concern and affective imagery, are also adopted to assess the environmental risk perception and to produce the quantitative representation of the same. Lastly, to examine the probability of migration from the Sundarban in comparison with nonmigrants, SLA model is taken into consideration and empirically tested with binary logistic regression. The detailed discussion of the above-mentioned techniques is illustrated in the respective chapters.

1.8 Organisation of the Book

This book is divided into nine chapters. Chapter 1 is the Introduction. It includes the problems of the study, literature gap, conceptual framework, objectives, data sources, sampling for primary data collection and methods. Chapter 2 describes the location, administrative setup and different demarcated areas of the biosphere reserve. It also discusses the universal importance of the Sundarban and its environmental condition with selected parameters. Chapter 3 deals with activities for means of living of the islanders and issues related to the same. Chapter 4 states the processes of migration partnering to stream and pattern, reasons of migration, work engagement before and after move and remittances flow. Chapter 5 deals with the environmental risk perception and its linkage to the migration. Chapter 6 depicts the various livelihood asset possessions by the households, and that ultimately portrays the development scenario of the Sundarban. Chapter 7 includes policies and institutional activities to promote the livelihood in the Sundarban. Chapter 8 is the empirical investigation of Sustainable Livelihood Approach (SLA) with binary logistic regression. Final Chapter 9 is the conclusion of the study, and some recommendations for the development of the Sundarban are also incorporated therein.

References

ADB. (2009). *Climate change and migration in Asia and Pacific: Executive summery*. Philippines: Asian Development Bank (ADB).

Aggarwal, M. (2015, March 18). Environment ministry revokes suspension of Goa mines. *The Mint*. http://www.livemint.com/Politics/FRAg9DGSJJv8vQYQS8fCLI/Environment-ministry-revokes-suspension-of-Goa-mines.html. Accessed 22 Oct 2016.

Banerjee, A. (1998). *Environment, population, and human settlements of Sundarban Delta*. New Delhi: Concept Publishing Company.

Banerjee, S., Gerlitz, J. Y., & Hoermann, B. (2011). *Labour migration as a response strategy to water hazards in the Hindu Kush-Himalayas*. Kathmandu: ICIMOD.

Belcher, J. C., & Bates, F. L. (1983). Aftermath of natural disasters: Coping through residential mobility. *Demography, 7*(2), 118–128.

Bhattacharya, S. (2007). *Lesson learn for vulnerability and adaptation assessment from India's first National communication*. Paper 7 of BASIC project, Winrock International. http://www.basic-project.net/data/final/Paper07India%20Lessons%20for%20VA%20from%20First%20National%20Communicat%85.pdf. Accessed 30 Sept 2016.

Black, R., Arnell, N. W., Adger, W. N., Thomas, D., & Geddes, T. (2012). Migration, immobility and displacement outcomes following extreme events. *Environmental Science and Policy, 27* (1), S32–S43.

Brooks, N. (2006, November). *Climate change, drought and pastoralism in the Sahel*. Discussion note for the word initiatives on sustainable Pastoralism. https://cmsdata.iucn.org/downloads/e_conference_discussion_note_for_the_world_initiative_on_sustainable_pastoralism_.pdf. Accessed 26 Nov 2016.

Brown, O. (2007). *Climate change and forced migration: Observations, projections and implications*. A Background Paper for the Human Development Report 2007/08. Geneva: United Nations Development Programme.

Camber, R., & Conway, G. R. (1992). *Sustainable rural livelihoods: Practical concept for 21st century* (IDS Discussion Paper 296). http://opendocs.ids.ac.uk/opendocs/bitstream/handle/123456789/775/Dp296.pdf?sequence=1. Accessed 25 July 2013.

Carling, J., Demba, P., Yassine, M., & Wu, J. (2013). *Migration aspirations in Senegal: Who wants to leave and why does it matter?* (European Policy Brief). http://www.eumagine.org/outputs/Policy%20Brief%20Senegal.pdf. Accessed 16 Sept 2015.

Connell, J., Dasgupta, B., Laishley, R., & Lipton, M. (1976). *Migration from rural areas: The evidence from village studies*. Delhi: Oxford University Press.

Czaika, M., & Vothknecht, M. (2012). *Migration as cause and consequence of aspiration* (Working Papers-57). International Migration Institute, University of Oxford. http://www.imi.ox.ac.uk/publications/wp-57-12. Accessed 16 Sept 2015.

Czaika, M., & Vothknecht, M. (2014). Migration and aspirations – Are migrants trapped on a hedonic treadmill? *IZA Journal of Migration, 3*(1), 1–21.

de Han, A. (1997). Rural-urban migration and poverty: The case study of India. *IDS Bulletin, 28*(2), 35–47.

DFID. (2001). *Sustainable livelihoods guidance sheets*. London: DFID.

DPD. (2009). *Human development report, South 24 Parganas*. Development and Planning Department (DPD), Government of West Bengal, Kolkata: HDRCC.

Du, B. L., & Zhang, J. W. (2000). Adaptation strategy for sea-level rise in vulnerable areas along China's coast. *Acta Oceanologica Sinica, 19*(1), 1–16.

Ericson, J. P., Vorosmarty, C. J., Dingman, S. L., Ward, L. G., & Meybeck, M. (2006). Effective sea-level rise and deltas: Causes of change and human dimension implications. *Global and Planetary Change, 50*(1-2), 63–82.

Fan, D. D., & Li, C. X. (2005). Complexities of China's coast in response to climate change. *Advance in Research on Climate Change, 1*(2), 111–114.

FAO. (2016). *World deforestation slows down as more forests are better managed*. http://www.fao.org/news/story/en/item/326911/icode/. Accessed 21 Oct 2016.

Ferries, E. (2014). *Disasters and displacement: What we know, what we don't know. Brookings*. https://www.brookings.edu/blog/planetpolicy/2014/06/09/disasters-and-displacement-what-we-know-what-we-dont-know/. Accessed 22 Oct 2016.

Findley, S. E. (1994). Does drought increase migration? A study of migration from rural Mali during the 1983–1985 drought. *The International Migration Review, 28*(3), 539–553.

Ghosh, A. (2012). *Living with changing climate: Impact, vulnerability and adoption challenges in Indian Sundarbans*. New Delhi: Centre for Science and Environment (CSE). http://cseindia.org/userfiles/Living%20with%20changing%20climate%20report%20low%20res.pdf. Accessed 20 Dec 2013.

Hazra, S., Samanta, K., Mukhopadhyay, A., & Akhandet, A. (2010). *Temporal change detection (2001–2008) study of Sundarban*. Final report. School of Oceanographic Studies, Jadavpur University. http://www.iczmpwb.org/main/pdf/ebooks/WWF_FinalReportPDF.pdf. Accessed 23 Jan 2013.

IOM. (2009). *Migration, environment and climate change: Assessing the evidence*. In F. Laczko, & C. Aghazarm (Eds). Geneva: IOM.

IOM. (2011). *Migration aspirations and experiences of egyptian youth*. Cairo: International Organization for Migration (IOM).

IPCC. (1992). *Climate change 1992: The supplementary report to the IPCC scientific assessment*. In J. T. Houghton, B. A. Callander, & S. K. Varney (Eds.). Cambridge: Cambridge University Press.

IPCC. (2000). *IPCC special report emission scenarios: Summary for policy makers*. A Special Report of Working Group III of the IPCC. Nairobi: IPCC. https://www.google.co.in/url?sa=t&rct=j&q=&esrc=s&source=web&cd=1&ved=0ahUKEwiIurLD7qrPAhUMQo8KHaQjAvcQFgggMAA&url=https%3A%2F%2Fwww.ipcc.ch%2Fpdf%2Fspecial-reports%2Fspm%2Fsres-en.pdf&usg=AFQjCNFa6spcdu6sJzKxafLiSblfImNCzw&sig2=NrjzZBPKqe0lxK3xDtW0vw&cad=rjt. Accessed 25 Sept 2016.

IPCC. (2007). *'Impacts, adaptation and vulnerability', Contribution of working group II to the fourth assessment report of the intergovernmental panel on climate change*. Cambridge: Cambridge University Press.

Iversen, V., Sen, K., Verschoor, A., & Dubey, A. (2009). Job recruitment networks and migration to cities in India. *The Journal of Development Studies, 45*(4), 522–543.

Jalais, A. (2011). *Forest of tigers: People, politics and environment in the Sundarban*. New Delhi: Routledge Taylor and Francis Groups.

Kanjilal, T., & Sarkar, K. (2009). আয়লা নয়, শেষের শুরু (It is not Aila, the Beginning of the End of Sundarban). Rangabalia, South 24 Parganas: Tagore Society for Rural Development.

Kanjilal, B., Mazumdar, P. G., Mukherjee, M., Mondal, S., Barman, D., Singh, S., & Mandal, A. (2013). *How healthy are the children of Indian Sundarbans?* The Sundarbans health watch report series: 1. Jaipur: IIHMR.

Khan, T. M. A., Singh, O., & Rahaman, M. S. (2000). Recent sea level and sea surface temperature trends along the Bangladesh coast in relation to the frequency of intense cyclone. *Marine Geodesy, 23*(2), 23–103.

Kothari, A., & Patel, A. (2006). *Environment and human rights: An introductory and essential readings*. New Delhi: National Human Rights Commission.

Laczko, F., & Aghazarm, C. (2009). Introduction and overview: Enhancing the knowledge base. In F. Laczko & C. Aghazarm (Eds.), *Migration, environment and climate change: Assessing the evidence*. Geneva: IOM.

Lal, M. (2001). Tropical cyclones in a warmer world. *Current Science, 80*(9), 1103–1104.

Lipton, M. (1980). Migration from rural areas of poor countries: The impact on rural productivity and income distribution. *World Development, 8*(1), 1–24.

Loucks, C., Barber-Meyer, S., Hossain, M., Barlow, A., & Chowdhury, R. (2010). Sea level rise and tigers: Predicted impacts to Bangladesh's Sundarbans mangroves. *Climatic Change, 98*(1-2), 291–298.

Massey, D. S. (1990). Social structure, household strategies, and the cumulative causation of migration. *Population Index, 56*(1), 3–26.

Massey, D., Axinn, W., & Ghimire, D. (2007). *Environmental change and out-migration: Evidence from Nepal*. Population Studies Center Research Report 07-615. http://www.psc.isr.umich.edu/pubs/pdf/rr07-615.pdf. Accessed 18 Nov 2016.

McDowell, C., & de Haan, A. (1997). *Migration and sustainable livelihoods: A critical review of the literature* (IDS Working Paper 65). http://www.ids.ac.uk/files/dmfile/Wp65.pdf. Accessed 21 Oct 2016.

McLeman, R., & Smith, B. (2006). Migration as an adaptation to climatic change. *Climatic Change, 76*(1), 31–53.

Mistri, A. (2012). *Natural resource dependence and migration from Sundarban Biosphere reserve: Case study of a remote Island*. The Centre for the Study of regional development, Jawaharlal Nehru University, (Unpublished Dissertation).

Mistri, A. (2013). Migration and sustainable livelihoods: A study from Sundarban biosphere reserve. *Asia Pacific Journal of Social Sciences, 5*(2), 76–102.

Mitra, A., Gangopadhyay, A., Dube, A., Schmidt, A. C. K., & Banerjee, K. (2009). Observed changes in water mass properties in the Indian Sundarbans (Northwestern Bay of Bengal) during 1980–2007. *Current Science, 97*(10), 1445–1452.

Mitra, A., Chowdhury, R., Sengupta, K., & Banerjee, K. (2010). Impact of salinity on mangroves of Indian Sundarbans. *Journal of Coastal Environment, 1*(1), 71–82.

MoEF. (2011). *The report of the Western Ghats ecology expert panel*. New Delhi: Ministry of Environment and Forests (MoEF), Government of India.

MoEF. (2013). *Reports of the high level working group on Western Ghats*. New Delhi: MoEF, Government of India.

Montgomery, J. D. (1991). Social networks and labor-market outcomes: Toward an economic analysis. *The American Economic Review, 81*(5), 1408–1418.

Munshi, K. (2003). Networks in the modern economy: Mexican migrants in the US labour market. *Quarterly Journal of Economics, 118*(2), 549–599.

Munshi, K., & Rosenzweig, M. R. (2006). Traditional institutions meet the modern world: Caste, gender and schooling choice in a globalizing economy. *American Economic Review, 96*(4), 1225–1252.

Muto, M. (2012). The impacts of mobile phones and personal networks on rural to-urban migration: Evidence from Uganda. *Journal of African Economies, 21*(5), 787–807.

Myers, N. (2005, May 23–27). Environmental refugees: An emergent security issue. In *13th Economic Forum*. Prague.

Perch-Nielsen, S. (2004). *Understanding the effect of climate change on human migration: The contribution of mathematical and conceptual models*. Department of Environmental Studies. Zurich: Swiss Federal Institute of Technology.

Raleigh, C., Jordan, L., & Salehyan, I. (2008). *Assessing the impact of climate change on migration and conflict* (The World Bank Group). Washington: The World Bank.

Reja, S. (2016). *Bengali construction workers in Kerala: A study on migration overriding distance*. Centre for the Study of Regional Development, Jawaharlal Nehru University (Unpublished Ph. D. thesis).

Scoones, I. (1998). *Sustainable rural livelihoods: A framework for analysis* (ISD Working Paper 72). https://www.staff.ncl.ac.uk/david.harvey/AEF806/Sconnes1998.pdf. Accessed 19 Oct 2016.

Serrat, O. (2008). *The sustainable Livelihoods approach*. Asian Development Bank. http://www.adb.org/sites/default/files/pub/2008/sustainable-livelihoods-approach.pdf. Accessed 12 Oct 2012.

Shivakoti, G. P., Axinn, W. G., Bhandari, P., & Chhetri, N. B. (1999). The impact of community context on land use in an agricultural society. *Population and Environment, 20*(3), 191–213.

Shurke, A. (1993). Pressure points: Environmental degradation, migration and conflict. In *Environmental change, population displacement, and acute conflict, June 1991*. Ottawa: Institute for Research on Public Policy. https://www.cmi.no/publications/file/1374-pressure-points-environmental-degradation.pdf. Accessed 22 Nov 2018.

Singh, O. P. (2007). Long-term trends in the frequency of severe cyclones of bay of Bengal: Observations and simulations. *Mausam, 58*(1), 59–66.

Singh, O. P., Khan, T. M. A., & Rahman, M. S. (2001). Has the frequency of intense tropical cyclones increased in the North Indian Ocean? *Current Science, 80*(4), 575–580.

Singh, A., Bhattacharya, P., Vyas, P., & Roy, S. (2010). Contribution of NTFPs in the livelihood of mangrove forest dwellers of Sundarban. *Journal of Human Ecology, 29*(3), 191–200.

Sjaastad, L. A. (1962). The costs and returns of human migration. *Journal of Political Economy, 70* (5), 80–93.

Small, C., & Nicholls, R. J. (2003). Aglobal analysis of human settlement in coastal zones. *Journal of Coastal Research, 19*(3), 584–599.

Stark, O. (1984). Rural-to-urban migration in LDCs: A relative deprivation approach. *Economic Development and Cultural Change, 32*(3), 475–486.

Stern, N. (2006). *The economics of climate change: Stern review*. Cambridge: Cambridge University Press.

TERI (1996). *Tata Energy Research Institute, New Delhi, Report No 93/GW/52*. Submitted to the ford foundation.

Tiffen, M., Mortimore, M., & Gichuki, F. (1994). *More people, less Erosion-environmental recovery in Kenya*. Chichester: Wiley.

Todaro, M. P. (1969). A model of labour migration and urban unemployment in less-developed countries. *The American Economic Review, 59*(1), 138–148.

UNDP. (2004). *Reducing disaster risk: A challenge for development, disaster reduction unit*. New York: Bureau for Crisis Prevention and Recovery.

UNDP. (2007). *Fighting climatic change: Human solidarity in a divided world, human development report 2007/2008*. Geneva: United Nations Development Programme.

Unnikrishnan, A. S., Kumar, K. R., Fernandes, S. E., Michael, G. S., & Patwardhan, S. K. (2006). Sea level changes along the Indian coast: Observations and projections. *Current Science, 90*(3), 362–368.

Wahba, J., & Zenou, Y. (2005). Density, social networks and job search methods: Theory and application to Egypt. *Journal of Development Economics, 78*(2), 443–473.

Wassmann, R., Nguyen, X. H., Chu, T. H., & Tuong, T. P. (2004). Sea-level rise affecting the Vietnamese Mekong Delta: Water elevation in the flood season and implications for Rice production. *Climatic Change, 66*(1), 89–107.

The Sundarban and Its Environment

2

Abstract

The Sundarban is the UNESCO designated World Biosphere Reserve (in 2001) and inscribed as the Natural World Heritage Site in 1987. It is the largest halophytic mangrove forest in a single block, famous for in-situ conservation of endangered endemics and magnificent biodiversity of flora and fauna. The mangrove forest and associates, marsh grass, phytoplankton, molluscs and other coastal vegetation in the Sundarban sequester a huge amount of 'Blue Carbon' that reduces the global warming. Nowadays, low-lying coastal areas and islands like the Sundarban are in the limelight of climate change. Sea level rise, sea surface temperature, both the intensity and frequency of tropical cyclones, air temperature, rainfall, salinity, surface water pH and water transparency, all the environmental indicators in the Sundarban, are increasing in trend. This chapter provides insights into the detailed examination of the environment in the Sundarban based on the existing literatures.

Keywords

Sundarban Biosphere Reserve (SBR) · Sundarban Tiger Reserve (STR) · Environmental change · Sea level rise · Tropical cyclone · Salinity

The Sundarban is one of the important UNESCO World Network of Biosphere Reserves in the Asia and Pacific region. It possesses outstanding universal values and unique features and is famous for its in-situ conservation. It plays a significant role for reducing the global warming. Moreover, it provides bread and butter of millions of islanders. The present chapter deals with the location and administrative setup of the Sundarban. The uniqueness and its universal values are also highlighted. At the last, the environment of the Sundarban is discussed in the light of climate change. Various studies on important environmental indicators are brought into the purview of discussion in detail.

A. Mistri, B. Das, *Environmental Change, Livelihood Issues and Migration*,
https://doi.org/10.1007/978-981-13-8735-7_2

2.1 Location and Designations

Sundarban Biosphere Reserve (SBR) extends between 21^0 32' and 22^0 40' north latitude and 88^0 05' to 89^0 00' east longitude, and it is demarcated by the river Hooghly to the west, the Bay of Bengal to the south, the Ichamati-Kalindi-Raimongal rivers to the east and the Dampier-Hodges line to the north (Map 2.1). The Sundarban had begun its journey as a Protected Forest (PF) in the year 1878; later it was declared as a Reserve Forest (RF) in 1928 under the Indian Forest Act, 1927. A part of the RF (2585 sq. km.) was declared as the Sundarban Tiger Reserve (STR) in 1973 under the programme, the Project Tiger (1973) which was launched according to the guideline of the Indian Wildlife (Protection) Act, 1972 (Table 2.1). Within the STR, a wildlife sanctuary, Sajnekhali (362.42 sq. km.) in 1976 and Sundarban National Park (1330.12 sq. km.) in 1984 were designed. In 1987, the National Park (NP) was inscribed as UNESCO's Natural World Heritage Site (Map 2.2). In 2007, a total 1699.62 sq. km. including NP was freshly designated as Core or Critical Tiger Habitat (CTH), and in 2009 the rest of the STR was notified as buffer zone. Within the buffer zone, 362.42 sq. km. (41%) is covered by Sajnekhali Wildlife Sanctuary, and the rest is the multiple-use zone, which is left for livelihood purpose. Apart from Sajnekhali Wildlife Sanctuary, another two wildlife sanctuaries, namely, Lothian Wildlife Sanctuary (38 sq. km.) and Haliday Wildlife Sanctuary (6.0 sq. km.), are demarcated out of the STR. However, 4263 sq. km. RF along with 5367 sq. km. the inhabited area, a total 9630 sq. km. was declared as the Sundarban Biosphere Reserve (Fig. 2.1) by the Government of India (GoI) in 1989, later on November 2001 by the UNESCO.

2.2 Administrative Setup

SBR extends over 102 islands (48 forested and 54 inhabited) and 19 Community Development Blocks (C. D. Blocks) of two districts, namely, South 24 Parganas (13 blocks) and North 24 Parganas (6 blocks), of the State of West Bengal (Map 2.3). Total 13 blocks in 4 subdivisions are under the administration (District Magistrates) of South 24 Parganas, and another 6 blocks within one subdivision are under the administration of North 24 Parganas (Table 2.2). Department of Sundarban Affairs (DoSA), Government of West Bengal, are entrusted for development activities in the backward area of the Sundarban. It formulates various plans for the improvement of basic infrastructure, livelihoods and socioeconomic condition as a whole. The Sundarban Development Board (SDB) is an apex body within the DoSA which acts as a directorate and coordinates between the bureaucrats and technocrats to the department activities, policy implementation, review and evaluation.

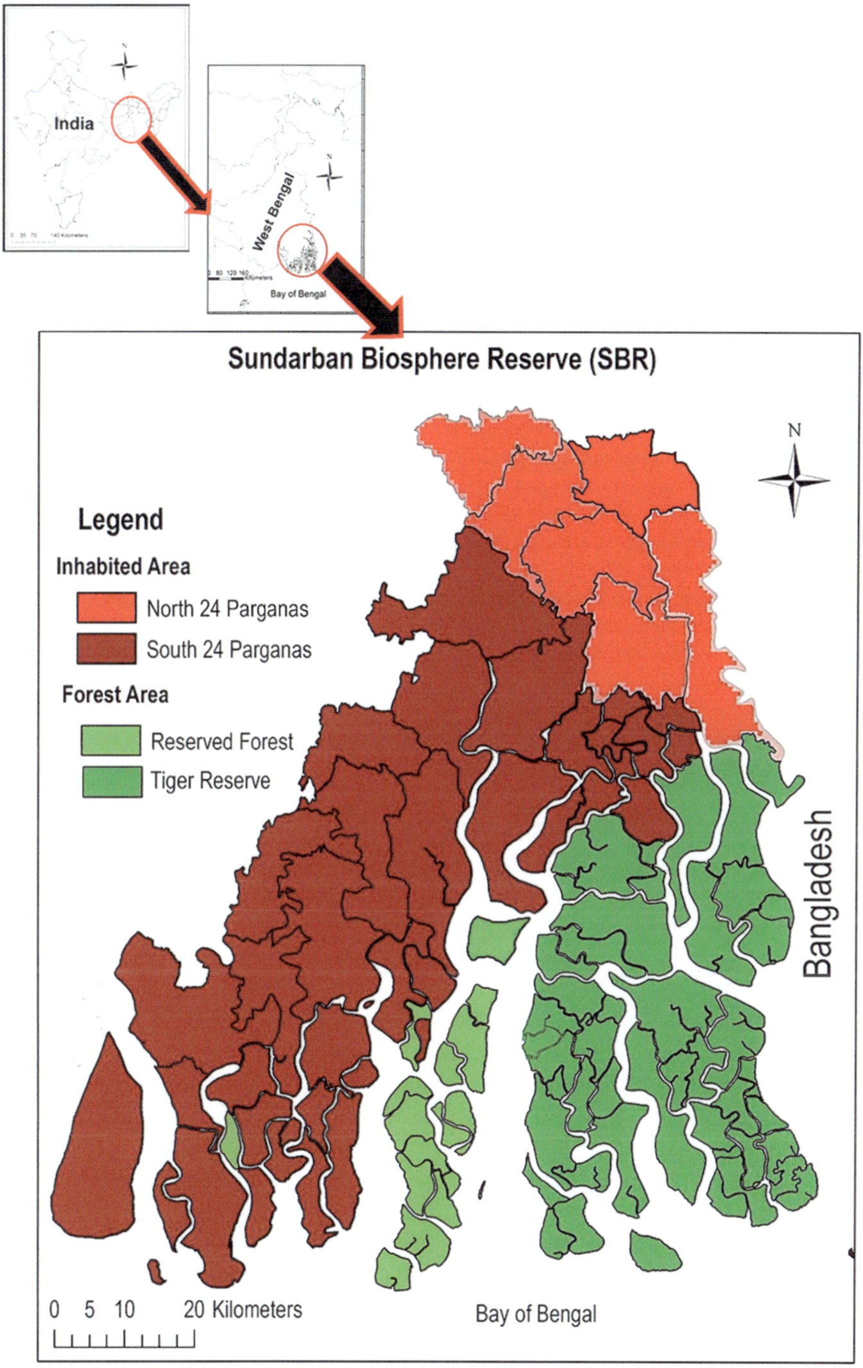

Map 2.1 Location map of the study area, Sundarban Biosphere Reserve, India (Source: Developed by authors)

Table 2.1 Demarcated areas and designations of Sundarban, India

Year	Legal designations	Area
1878	The area was declared as Protected Forest (PF)	–
1928	It was declared as a Reserve Forest (RF)	4263 sq. km
1973	One of the first nine Sundarban Tiger Reserves (STR) declared under the Project Tiger Scheme	2584.89 sq. km covering land area 1600 sq. km and water body 985 sq. km
1976	A part of the Sundarban Tiger Reserve (STR) was notified as the Sajnekhali Wildlife Sanctuary	362.42 sq. km
1984	The core area of the tiger reserve was declared as the Sundarban National Park (NP)	NP is 1330.12 sq. km, of which 124.40 sq. km is preserved as a primitive zone to act as a gene pool
1987	The National Park was inscribed as a Natural World Heritage Site by UNESCO	1330.10 sq. km
1989	The STR along with the adjoining forest area and human habitat were declared as a Sundarban Biosphere Reserve (SBR) by the Government of India	9630 sq. km (4263 sq. km. forest area + 5367 sq. km inhabited area)
2001	The SBR was enlisted on UNESCO World Network of Biosphere Reserves (WNBR) under the Man and Biosphere (MAB) programme	9630 sq. km
2007	Freshly designated core of the STR or Critical Tiger Habitat (CTH)	1699.62 sq. km
2009	Buffer Zone notification	885.27 sq. km in including 362.42 Sajnekhali Wildlife Sanctuary

Source: Authors' collection from different official documents

The forest part of the SBR includes 2584.89 sq. km. STR, and the rest are the non-STR or divisional RF. The territory of the STR is divided into four 'ranges', and each range is divided into several 'beats' (Table 2.3). There are 14 beats within 4 ranges in STR.

The STR is headed by a Field Director whose rank is equivalent to the Chief Conservator of Forest (Fig. 2.2). The Field Director is supported by one Deputy Field Director, two Assistant Field Directors, Range Officers, Beat Officers and other Frontline Staffs. Apart from STR, the rest of the RF is under a Joint Director. He also works under the Field Director of SBR and supported by two separate Divisional Forest Officers for the forest under two districts. The maximum proportion of the divisional forest is under South 24 Parganas. Hence, the Divisional Forest Officer of South 24 Parganas is assisted by two Assistant Divisional Forest Officers. Apart from that, the Sundarban shares its border with Bangladesh. The border-related issues are handled by the Border Security Force (BSF), Ministry of Home Affairs.

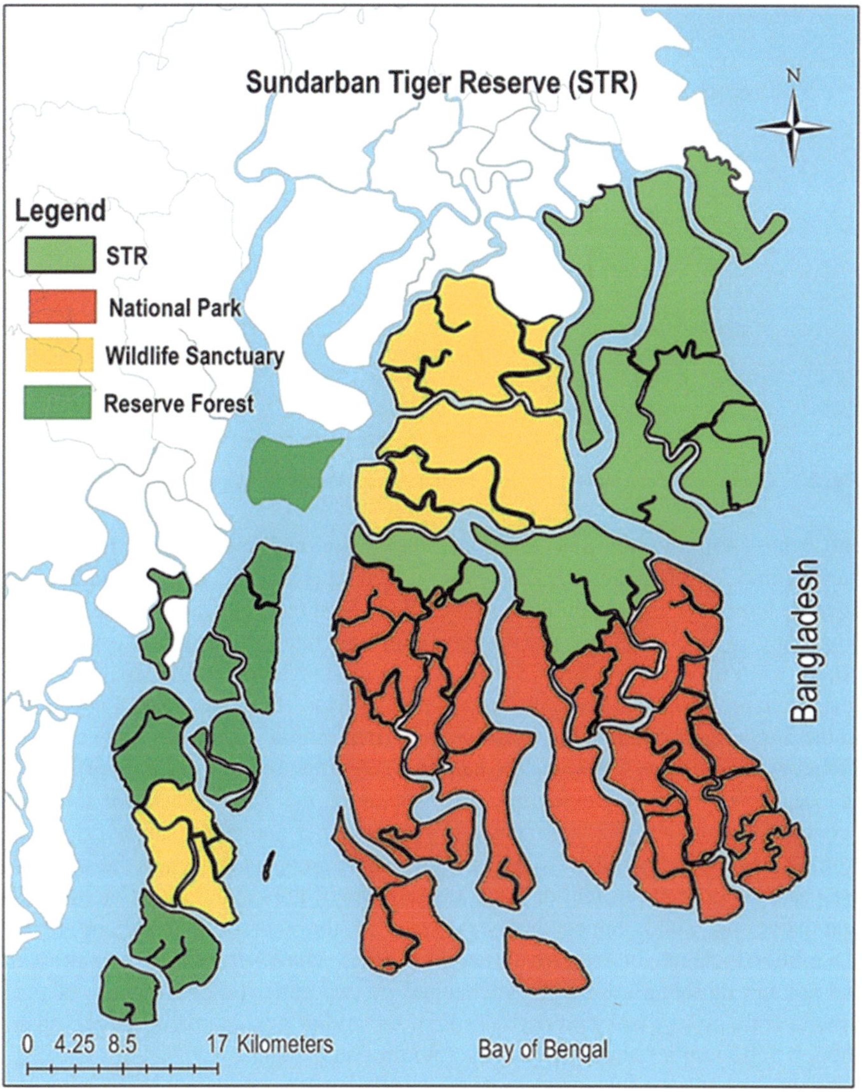

Map 2.2 Sundarban Tiger Reserve (STR) and its Divisions in India (Source: Developed by authors)

2.3 Importance of the Sundarban

2.3.1 The Sundarban: Outstanding Universal Value

The Sundarban is the world's largest delta constituted by the confluence of the Ganges and Brahmaputra rivers. The intricate network of interconnecting waterways and creeks, mudflats or tidal flats, swampy land, low-lying islands, sandy beaches

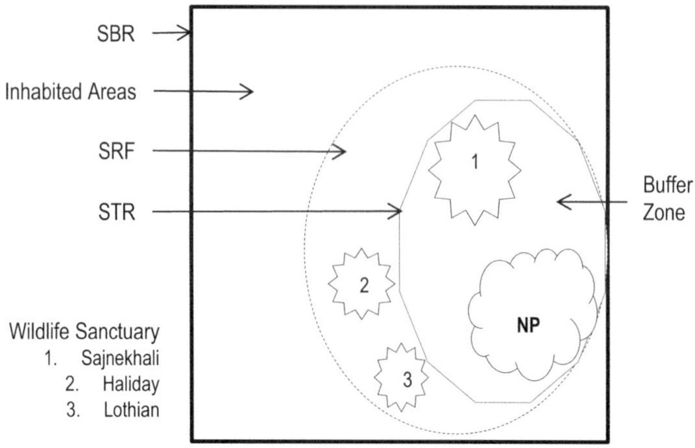

Fig. 2.1 Graphical representation of SBR, India (Source: Developed by authors)

and dunes, semi-diurnal tidal flow, ongoing erosion and deposition processes and various species of mangrove forests along with habitats of several hundreds of aquatic, terrestrial and amphibian are the salient features of the Sundarban. It is the single largest block of halophytic mangrove forest in the world extended over around 10,200 sq. km. across India (4200 sq. km.) and Bangladesh (6000 sq. km.). The mangrove forest, marsh grass, phytoplankton, molluscs and other coastal vegetation in the Sundarban contain a huge amount of 'Blue Carbon' that reduces the effect of global warming. Blue Carbon is the carbon (CO_2) captured by the living organisms of oceans and coastal ecosystems. The islands of the Sundarban have immense economic importance as it acts as a barrier to cyclones and storm surges originated from the Bay of Bengal, shore stabiliser and sediment trap and provides the shelter as well as a supplier of nutrient of the wide varieties of flora and fauna. Catching fish and gathering wood, honey and wax serve as the source of livelihood to the 0.5 million local inhabitants (Mistri and Das 2015). Moreover, every year a substantial amount of foreign money is exchanged by exporting the high quality of tiger prawn, crab, honey, wax and dry fish. Ecotourism has a wide potentiality to create the job opportunities and flow revenue into the Sundarban.

2.3.2 Significant Ongoing Ecological and Biological Processes

Indian Sundarban meets the two criterions, such as (ix) and (x), out of ten prescribed by the UNESCO to be included on the World Heritage list (Appendix 1). World Heritage sites are selected on the basis of six cultural criterions [(i) to (vi)] and four natural criterions [(vii) to (x)]. Significant ongoing ecological and biological processes are covered under criterion (ix). The Sundarban is an outstanding example of ongoing ecological and biological processes in the evolution and development of terrestrial, coastal and marine ecosystems and communities of flora and fauna. The

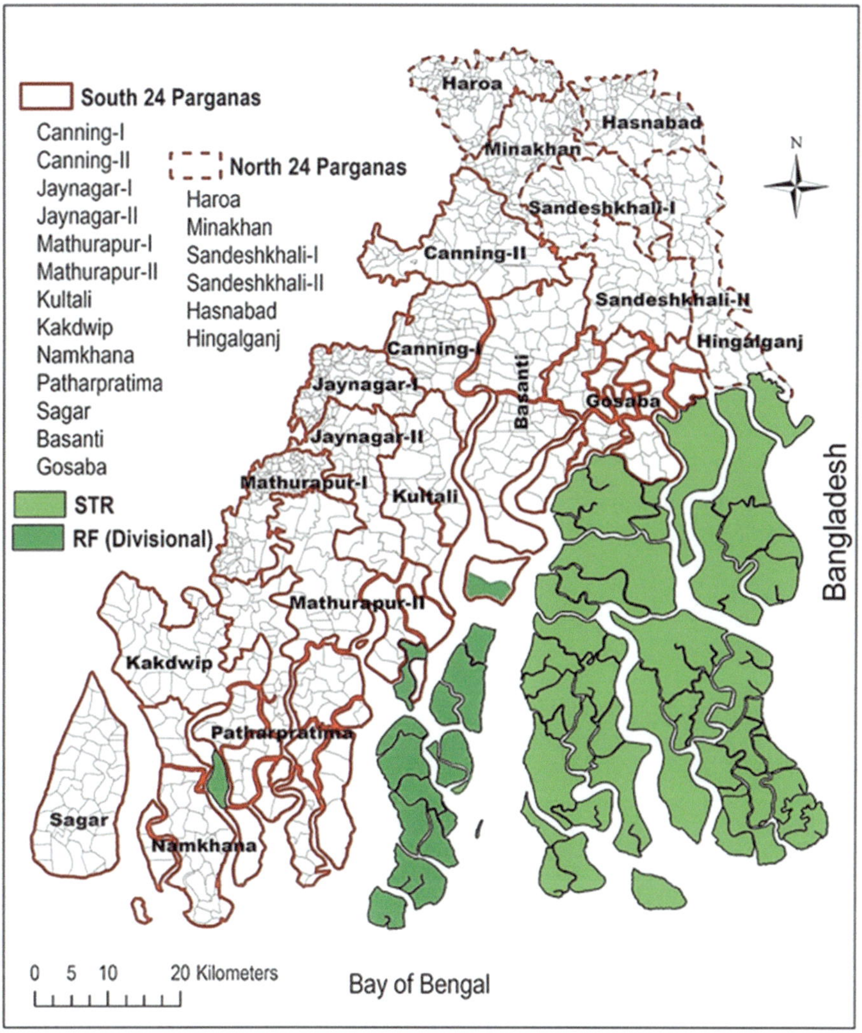

Map 2.3 Administrative set of Sundarban Biosphere Reserve, India (Source: Developed by authors)

islands are continuously changing its shape and size and forming new ones by the action of tidal erosion, river current, storm surge and ocean waves and the deposition of alluvium transported by the great rivers, the Ganges and Brahmaputra. These processes are associated with high temperature and humidity, cyclone, heavy annual rainfall (150–200 cm), floods and plant colonisation. Low-lying islands and swampy areas act as wetland nurseries for different organisms, mangroves, fish, crab and various types of prawn. Brackish water in the estuaries and riverine environment are favourable breeding ground for some of the marine species those seasonally migrate for the same. The islands with dense mangrove forest act as a climate buffer against storm surge, high tide and flash flood.

Table 2.2 Administrative units, inhabited areas of Sundarban, India

Districts	Subdivisions	C. D. Blocks
South 24 Parganas	Kakdwip	Sagar, Namkhana, Kakdwip, Patharpratima
	Diamond Harbour	Mathurapur-I, Mathurapur-II
	Baruipur	Kultali, Jaynagar-I, Jaynagar-II
	Canning	Canning-I, Canning-II, Basanti, Gosaba
North 24 Parganas	Basirhat	Hingalganj, Hasnabad, Sandeshkhali-I, Sandeshkhali-II, Haroa, Minakhan.
Total	5 (4 + 1)	19 (13 + 6)

Source: South & North 24 Parganas Districts Handbooks, B.A.E. & S (2011)

Table 2.3 Administrative units of STR in India

Ranges	Beats
Basirhat range	Bagna, Burirdabri, Jhingekhali, Khatuajhuri and Harinbhanga (in Harikhali)
Sajnekhali Wildlife Sanctuary	Dobanki, Sajnekhali and Duttar
National Park (east)	Chamta, Baghmara, Chandkhali (new)
National Park (west)	Haldibari, Netidhopani, Kendo

Source: STR (2015)

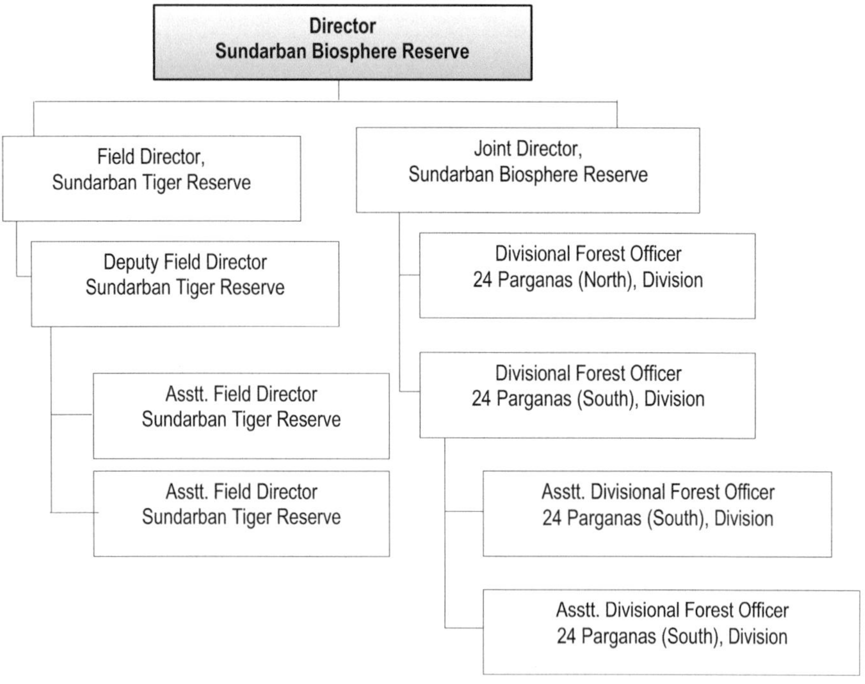

Fig. 2.2 Administration of SBR (Forest part) (Source: Developed by authors)

2.3.3 Significant Natural Habitat for In-situ Conservation of Biological Diversity

The existence of a large amount of significant natural habitat which needs to be conserved to maintain biological diversity is covered in criterion (x). The Sundarban is famous for its magnificent biodiversity, both the terrestrial and marine environments. More than 220 species of birds, 250 species of fish, 56 species of reptiles, 40 species of mammals and 84 species of mangroves and associates are found in Indian Sundarban.[1] It is also the natural habitat of globally endangered species, namely, the Royal Bengal Tiger, Irrawaddy and Ganges River Dolphins, Estuarine Crocodiles and the critically endangered endemic[2] River Terrapin (*Batagur Baska*), and King Cobra is on the vulnerable list (Feature Box 2.1). Two out of four species of primitive horseshoe crab are found in the Sundarban. Some species were extinct from the Sundarban such as Javan rhinoceros, water buffalo and Indian muntjac. The Royal Bengal Tiger is the primary attraction of the Sundarban, and it is the only mangrove habitat in the world for the endangered cat species. Around 76 tigers are estimated in Indian Sundarban in 2014 (Jhala et al. 2015). Pakhiralaya near the Sajnekhali Wildlife Sanctuary, Gosaba block and Bakkhali in Namkhana block are well-known for waterfowl and migratory birds.

Feature Box 2.1: IUCN Red List

The International Union for Conservation of Nature (IUCN) Red List of threatened species/IUCN Red List/Red Data List established in 1964. It is the world's most comprehensive inventory of the global conservation status of biological species. The IUCN is the world's chief authority on the conservation status of species. A series of Regional Red Lists are prepared by countries or organisations on the basis of the risk of extinction of species within a management unit. IUCN Red List classifies the species into nine groups.

In the Sundarban, the Royal Bengal Tiger, Irrawaddy and Ganges River Dolphins and Estuarine Crocodiles are globally endangered (EN) species. River Terrapin (*Batagur Baska*) is the critically endangered (CR) endemic, and King Cobra is included in the vulnerable category (VU). Javan rhinoceros, water buffalo and Indian muntjac were extinct (EX) from the Sundarban.

(continued)

[1] Mangrove Interpretation Center, Sajnekhali Beat Office, Sundarban Tiger Reserve, 2015.

[2] Endemic species are geographically constrained to one particular place on the earth. These are more likely to form in biologically isolated areas such as islands and large bodies of water and mountain areas. Endemic species run a higher risk of extinction because of their geographic isolation.

Feature Box 2.1 (continued)

- Extinct (EX)
- Extinct in the wild (EW)
- Critically endangered (CR)
- Endangered (EN)
- Vulnerable (VU)
- Near threatened (NT)
- Least concern (LC)
- Data deficient (DD)
- Not evaluated (NE)

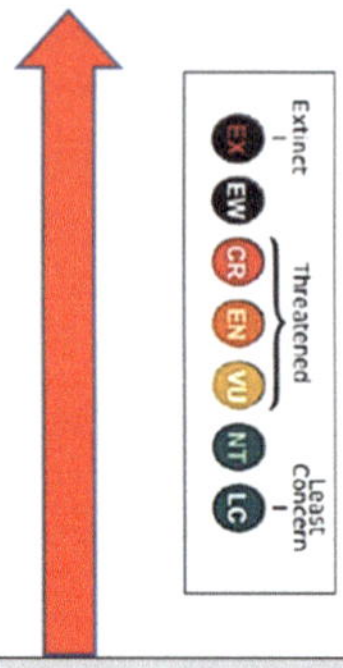

2.3.4 Sundarban Biosphere Reserve (SBR): An Integrity

It has already been mentioned that nearly 42%, equivalent to 4263 sq. km., of total 10,200 sq. km. of the Sundarban (Indian and Bangladesh) is under Indian territory. Within the 4263 sq. km. forested area, only 1330.12 sq. km. is the National Park (NP) inscribed as Natural World Heritage Site by UNESCO in 1987. The 4263 sq. km. forested area along with adjacent 5367 sq. km. inhabited area, together with 9630 sq. km., has been designated as UNESCO's Biosphere Reserve under the Man and Biosphere (MAB) programme in November 2001. The SBR comprises 102 islands, 54 inhabited islands and 48 forested islands. Every UNESCO's Biosphere Reserve is designed with three distinctive segments, namely, core, buffer and transition zones. Inhabited area falls under the transition zone and in between core and transition are buffer zone (Fig. 2.1). Buffer is allowed for fishing and other livelihood activities with proper licence and permit. The core areas are strictly prohibited for human interference. This united formation is conducive for better conservation of biological diversity, to ensure environmental sustainability and to enhance people's livelihood. Moreover, Biosphere Reserve promotes participatory approach to the conservation of an area by involving local communities and NGOs at a large scale. It also helps to draw global attention and additional funding for the development.

2.4 Environment of the Sundarban

Nowadays, low-lying areas and islands like the Sundarban are at the top of the list of global warming due to sea level rise, coastal erosion, extreme climatic events and ecosystem loss at very high confidence (IPCC 2007a). At the end of the 2000s, IPCC (2007a) expressed their concern on the effect of climate change in the Sundarban. The environment of the Sundarban is discussed with the help of existing literature below.

2.4.1 Sea Level Rise (SLR)

The absolute height of the Sundarban varies in between 0.90 and 2.11 metres above the mean sea level (MSL) (DPD 2009). The sea level of the Sundarban has observed change over the years. A study by Nandy and Bandyopadhyay (2011) on the observation[3] of four stations, namely, Sagra, Gangra, Haldia and Diamond Harbour, of the Hooghly estuaries has revealed that the rate of sea level changes for the stations are −3.82, +0.89, +2.43 and + 4.85 mm per year, respectively, during the reference period 1937–2006. The study (Hazra et al. 2002) based on the tidal record of Sagar Island has examined a rise in MSL at 3.14 mm per year during[4] 1985–1998. In an observation from Hiron Point in the Bangladesh Sundarban, it has been found that the trend of SLR is 4.0 mm per year during 1977–1998 and it is caused mainly by increasing of monsoonal rain and sea surface temperature (SST) (Singh 2002). However, from the studies, sea level rise in the Sundarban can be estimated 3.14–4.0 mm per year. It is far higher than all India average 1.0 mm per year (Unnikrishnan et al. 2006) and the global average that is 1.8 ± 0.5 mm per year during 1961–2003 or 1.7 ± 0.5 mm per year for the twentieth century (Bindoff et al. 2007). The SLR in Bangladesh Sundarban is attributed by the monsoonal rain, SST and somehow land subsidence (Singh 2002; Khan et al. 2000), but the SLR for Indian Sundarban is not pinpointed as of yet. The SST, river discharged by the Hooghly and other upper streams, coastal erosion and accretion of suspended sediments are differently associated with the SLR in different parts of Indian Sundarban (Nandy and Bandyopadhyay 2011; Hazra et al. 2002).

2.4.2 Surface Water Temperature

The average annual surface water temperature in the Sundarban is 28°–29 °C. A study by Mitra et al. (2009) has examined that surface water temperature in the Sundarban has increased at the rate of 0.05 °C per year in the past 27 years, 1980–2007. The study has also observed that there are regional variations of surface water temperature between eastern and western parts of the Sundarban during the pre-monsoon and post-monsoon period. The water temperature in western Sundarban has risen by 6.14%, where the eastern part has registered an increase of about 6.12% during that period (1980–2007). Another study during a very short span, 2001–2008, has estimated that the water temperature in the Sundarban is rising at the rate of 0.0453 °C per year (Hazra et al. 2010). A study on Bangladesh Sundarban, estuarine water near Hiron Point, has registered an increasing trend of around 1.0 °C (0.071 °C per year) in May and 0.5 °C (0.036 °C per year) in November during 1985–1998 (14 years) (Khan et al. 2000). From these studies, it can be suggested that the surface water temperature in the Sundarban is rising at the

[3]The study is based on the data sets of Permanent Service for Mean Sea Level (PSMSL).

[4]The study used tide gauge data of five discontinuous years within a span of 14 years, such as 1985, 1990, 1996, 1997 and 1998.

rate of around 0.04^0–0.05 °C per year. This trend is higher than the global average that is 0.006 °C per year, even higher than the Indian Ocean average 0.02 °C per year (IPCC 2007b).

2.4.3 Tropical Cyclone

About five to six cyclones in a year form in the Bay of Bengal and Arabian Sea. These are more frequent[5] in the Bay of Bengal than the Arabian Sea. After the analysis of more than 100 years data sets, it has been observed that most of the cyclone form on May, October and November; among these October and November are known for the most severe cyclonic storm (SCS) occurrence month (Singh et al. 2001; Khan et al. 2000). The frequency of tropical cyclonic storm over the Northern Indian Ocean has twofold increased during November in the last 122 years (1877–1998) (Singh et al. 2001). The intensification rate of tropical disturbances to be reached at the cyclonic stage has significantly increased by 17% (Singh et al. 2001) and 26% increase to the SCS stage in the Bay of Bengal during November in the past 129 years, 1877–2005, (Singh 2007). There is a strong association between SST and cyclogenesis. Tropical storms become cyclone if the sea surface temperature is more than 26 °C (Khan et al. 2000). The increasing trend of the frequency and intensification rate of the cyclonic storm during intense cyclone months (May, October and November) in the Bay of Bengal are significantly correlated with the rise of surface water temperature (Khan et al. 2000).

Table 2.4 portraits the list of tropical cyclones in the northern Bay of Bengal, during 1999–2018. In the last 10 years (2005–2015), on an average, one tropical cyclone in a year was formed over the northern Bay of Bengal which includes three of each super cyclonic storms, very SCSs and SCSs. Among them, Aila in 2009 severely ravaged the life of the Sundarban. Its velocity was 120 km per hour with 10–15 metre water surge. More than eight lakh people were displaced, and the death toll was reached to nearly 120 people (Mistri 2011). The estimated financial loss surpassed the cost of any cyclonic disasters occurred during the past two decades (1980s and 1990s) in the Sundarban. About 31,000 hectare of cultivated land was unfertile due to high salinisation that leads to aftermath 2–3 years complete crop failure, and approximately Rs. 100 core crop was lost (Mistri 2011). Apart from Aila, Sidr in 2007, Bijli in 2009 and Phailin in 2013 also affected the life of the Sundarban people.

2.4.4 Air Temperature

The mean annual surface air temperature in India has increased by 0.4 °–0.6 °C in the last 10 years (Bhattacharya 2007; Rupa Kumar et al. 2006). It widely affects the

[5]The ratio of frequency between Bay of Bengal and Arabian Sea is 4:1.

Table 2.4 Tropical cyclone in northern Bay of Bengal during 1999–2018

Name	Date of occurrences	Speed (knots) (1 knot = 1.852 km/h)	Category
–	28 October 1999	>140	Super cyclonic storm
–	28 October 2000	<40	Cyclonic storm
–	19 May 2003	<60	SCS
–	17 May 2004	<60	SCS
–	2 October 2005	<40	Cyclonic storm
Mala	24 April 2006	>120	Super cyclonic storm
Not given	13 May 2007	<60	SCS
Sidr	15 November 2007	>120	Super cyclonic storm
Not given	28 June 2007	>120	Super cyclonic storm
Rashmi	26 October 2008	>40	Cyclonic storm
Nargis	27 April 2008	<120	Very SCS
Bijli	16 April 2009	<60	SCS
Aila	24 May 2009	<60	SCS
Giri	19 October 2010	<120	Very SCS
Phailin	4 October 2013	<120	Very SCS
Komen	31 July 2015	<40	Cyclonic storm
Titli	11October 2018	<119	Very severe cyclonic storm

Note: IMD Classification of low pressure: <17Knots (<32 Kmph), low-pressure area; 17–27 (32–50), depression; 28–33 (51–59), deep depression; 34–47 (60–90), cyclonic storm; 48–63 (90–119), severe cyclonic storm (SCS); 64–119 (119–220), very severe cyclonic storm; >119 (>220), super cyclonic storm
Source: Hazra et al. (2010, p. 58)

yield rate of crops in different parts of India. In every 0.5 °C increase of temperature in winter is estimated to reduce the yield rate of rainfed wheat by 0.45 tonne per hectare in India (Lal et al. 1998). Likewise, a 2.0 °C increase in mean air temperature is projected to decline the rice yield by 0.75 tonne per hectare in the high-yield areas and by 0.06 tonne per hectare in the low-yield coastal areas (Bhattacharya 2007). In the Sundarban, surface air temperature has increased at the rate of 0.1058 °C per year during 2002–2009 (Hazra et al. 2010). But the previous study by the same team (2002) estimated that the increasing rate was 0.019 °C per year during 1970–2000. However, the current estimation by the team (2010) shows an alarming increasing of air temperature around 1.0 °C per decade. They argue that this anomaly may be due to the formation of a local heat pool that is the setup of industrial hubs near to the Sundarban or the expansion of Kolkata Urban Agglomeration (UA).

2.4.5 Rainfall

India highly relies on rainfed agriculture, mainly monsoonal rain. Nearly 60% croplands are rainfed. The summer monsoon rainfall is projected to rise nearly 20% in all over India, except Punjab, Rajasthan in the northwest and Tamil Nadu in the south (Bhattacharya 2007). The intensity of rainfall is also estimated to increase by 1–4 mm per day in almost all over India, except some pockets in Northwest India where it may decrease at the rate of 1.0 mm per day. But it is estimated that there will be a decline of overall monsoon break days in India. The Western and Central India accounts for declining of more than 15 days. In contrast, foothills of the Himalayas and Northeast India may gain extra 5–10 monsoon breaking days. Like entire India, the monsoon rain has significantly increased in the Sundarban during 2001–2008 at the rate of 0.0041 mm per hour (Hazra et al. 2010). Average annual rainfall in the Sundarban is 1920 mm. But the rainfall is erratic in nature and full of uncertainty (Chand et al. 2012).

2.4.6 Salinity: Water and Soil

Rising salinity in water and soil is the major threats to the mangrove forests and livelihood of Sundarban people. Mangroves are not salt lovers; rather these are salt tolerant (Mitra et al. 2010). The excessive saline conditions retard germination, stunt growth and development of mangrove species. Mangroves can thrive at a range of salinity level 4–28 psu.[6] It depends on species to species. The level of water salinity in the Sundarban varies from minimum 8.66 psu in post-monsoon to maximum 26.59 psu in pre-monsoon (Mitra et al. 2010). It changes seasonally like pre-monsoon> post-monsoon> monsoon, and it also varies region wise, such as the western Sundarban is always recorded lower salinity compared to the eastern Sundarban. The average salinity difference between pre-monsoon and post-monsoon period is almost 5.0 psu per year (Mitra et al. 2010). During the past 29 years' observation (1980–2007), it has found that western Sundarban has registered a continuous decrease of salinity at the rate of 1.67 psu per decade, whereas eastern Sundarban has increased at the rate 2.0 psu per decade. The difference in salinity condition between the western and eastern section of the Sundarban is attributed by their geographical settings (Mitra et al. 2009). The rivers in the western Sundarban (Hooghly and Muriganga) are the continuation of the river Ganges. Hence there is a continuous supply of the snowmelt water from the Himalaya which reduces the salinity level. Due to global warming, the Himalayan Glacier is receding rapidly at the rate of 23 m per year during the previous 200 years (Mitra et al. 2009). On the contrary, clogging of the connection of the estuaries with the main channel of the

[6]Practical salinity units (psu).

Ganges due to heavy siltation and solid waste disposal from megacity Kolkata has reduced the fresh water supply, and intrusion of the saline water via tidal flow from the Bay of Bengal has been increasing the surface water salinity in eastern Sundarban. However, owing to the increase of salinity, some mangrove species, such as *Heritiera fomes* (*Sundari*) and *Nypa fruticans* (*Golpata*), are declining rapidly, especially in central and eastern Sundarban. *Nypa fruticans* is widely used as roof and fetching material of temporary houses of the local inhabitant. Due to rapid declination, the collection of *Nypa fruticans* has also been strictly prohibited by the SBR authority.

The Sundarban is a traditional rice cultivating area, and agriculture is mostly rainfed that is practised in the monsoon. Soil salinity beyond 6.0 ppt[7] hampers that rice cultivation. The northern part of the Sundarban has witnessed nearly 8.0 ppt of soil salinity, whereas the southern part, which is closer to the Bay of Bengal, has registered 8.0–20.0 ppt that is much higher than the safe limit (4.0–6.0 ppt) for rice cultivation (Ghosh 2012). Soil test funded by the European Commission for Humanitarian Aid (ECHA) examined that soil salinity had increased more or less 5.0 feet below top soil which led to nearly 2–3 years complete crop failure, after the SCS, Aila in 2009. Meanwhile, the level of salinity in the inland fresh water storages for irrigation was increased rapidly aftermath of Aila. The pond and canal salinity was raised up to 30 ppt, and shallow dip tube well water was raised to 28 ppt (Goshthakur 2009) that severely hampered the fisheries as well as farming.

2.4.7 Surface Water pH

Surface water pH is the indicator of the atmospheric CO_2 budget. Excessive CO_2 in the atmosphere dissolves into the ocean in an enormous amount and makes the water more acidic through lowering the pH condition that negatively affects the marine ecosystem (Doney and Levine 2006). The observed range of pH in the Sundarban is 8.25–8.33 during 1980–2007, which is higher than the global average of 8.17 (Mitra et al. 2009). The western side of the Sundarban has recorded a continuous decrease of pH at the rate of 0.015 per decade during that period which may be an indication of the direct absorption of excess CO_2 into the surface water. Meanwhile, in the eastern side, the pH level has very slightly been increasing over the years, but there is a lack of consistency.

2.4.8 Water Transparency

Water transparency in the Sundarban has changed over the year. Erosion and accretion along with churning action increase the saturation of suspended solids that reduce the transparency of water in the Sundarban. Low water transparency

[7]Parts per thousand (ppt).

affects the growth and survival of phytoplankton by reducing sunlight penetration into the water, and this ultimately declines the fish production. Water transparency in the Sundarban has declined from 7.0 cm to 9.0 cm at the rate of 2.3 cm per decade during 1980–2007 (Mitra et al. 2009), and eastern Sundarban is more transparent than the western part.

2.4.9 Accretion and Erosion

The Sundarban is an active delta, where building process is continuing through accretion and erosion processes. Studies (Ghosh 2012; Hazra et al. 2010; Rudra 2010) reveal that there has been no significant gain of land during the last 80 years (1930–2008) but the net loss has doubled during the last 8 years, 2001–2008. During 1930–2000, the net loss was 2.85 sq. km. per year, and it has increased to 5.5 sq. km. per year during 2000–2008.

Western Sundarban is severely affected by the erosional processes compared to the eastern Sundarban. Increasing flow of freshwater in the Ganga-Bhagirathi-Hooghly river due to the Himalayan deglaciation (Mitra et al. 2009) and strengthened velocity of the Hooghly channel due to construction of the Farakka Barrage on the Ganges and some incomplete and abandoned construction work of the Port Trust of India (Ghosh 2012) led massive erosion in western part of the Sundarban. Apart from land erosion, declining channel depth due to the high rate of deposition reduces the channel carrying capacity. During the high tide period or in rainy season, excess water of channels' carrying capacity spills over the mud-built embankment and claims the inhabited land breaking of the same. Every year a substantial proportion of cultivable land has been sacrificed by the river grasp (Feature Box 2.2).

2.4.10 Land Use and Land Cover Change

There are evidences of rapid change of land use and land cover in the Sundarban. Though the dense forest cover has not reduced significantly, the degraded forest area has reduced to 332.01 sq. km. from 404.89 sq. km. in between 2001 and 2009 (Table 2.5). The decline of total forest area is mostly attributed by the coastal erosion in the forested islands. The area of saline blanks has increased to 74.80 sq. km. from 38.93 sq. km. during 2001–2009.

In the inhabited islands, settlement density pattern has changed over the years. In the 2001 Census, the total population was 3.8 million, which has increased to 4.43 million in the 2011 Census at the rate of 1.65% per year (geometric growth), whereas the state (WB) average annual growth rate is 1.31% per year. The study by Hazra et al. (2010) reveals that the area under agricultural land has been converted into the settlement, vegetation and aquaculture over the years (2001–2009). In South

Feature Box 2.2: *'Ring Bandh'*

'Ring Bandh' is a local term which refers to the newly constructed arching shaped mud-built embankment in the Sundarban. After breakdown of an embankment, a new one is constructed away from the previous position. The landward erosion and extreme river current erode the base of the broken embankment and try to merge with the riverbed. To construct a *ring bandh*, a substantial amount of land has to leave for river grasp result in certain number of households is displaced. Plate 2.1 portraits the broken embankment wiped out by the water surge of Aila in 2009, and Plate 2.2 shows *'ring bandh'* erected after Aila leaving some amount of agricultural land to the river grasp.

Plate 2.1 The broken embankment by the storm surge of SCS, Aila in 2009

Plate 2.2 'Ring Bandh' after Aila

24 Parganas, the conversion of farmland into the settlement and vegetation is the predominant character of the land use and land cover changes, whereas, in North 24 Parganas, the conversion of farmland into the aquaculture farms and most recently to the brick kiln industries is the major shift.

Table 2.5 Land use and land cover (sq. km.) change in SBR during 2001–2009

Sl. no	Land use/land cover classes	Year 2001	Year 2009
1	Dense forest	1655.878	1651.3275
2	Degraded forest	404.887	332.0008
3	Saline blanks	38.93	74.7965
4	Settlement with vegetation	1226.334	1666.43
5	Agricultural land	2149.615	1691.246
6	Aquaculture farm	603.603	649.1
7	Water body	232.888	250.6531
8	Mudflats	23.897	12.6135
9	Sand (beaches/dunes)	8.0835	8.7664
10	Reclaimed land from forest	14.512	12.644
11	Swamp	14.847	20.41
	Total	**6373.4745**	**6369.9878**

Source: Hazra et al. (2010, p. 18)

All the above selected environmental indicators in the Sundarban have been changing over the years. Nowadays, climate or environmental change is a serious debate due to data limitation (timescales) to prove the rate of changes of parameters as evidence. The Sundarban is also not an exception. To examine climate change, it needs 'long term continuous change (increase or decrease) to average weather conditions or the range of weather' (Dinse 2012, p. 1). In climatology, a 'change' is defined as a change in average weather conditions or the range of weather over base time periods. Environmentalists roughly assume a minimum base period as 30 years. Until now, whatever data sets have been used in support of climate change in the Sundarban, most of these are not more than a climatological normal (30 years). Moreover, most of the environmental indicators have recorded very minimal change and are highly influenced by local factors; even, some of the indicators have been recorded positive (good) change. Therefore, it is argued that whatever changes in the variables have been observed could be defined as climate variability instead of climate change. There is nothing to be worried about the present trend of change in the Sundarban. It may be equilibrated through its dynamic mechanism within the system in the near future. However, whether 'climatic change' or 'climatic variation' is observed in the Sundarban, the argument is beyond of the scope of the study. The condition of environmental parameters over the years indicates the gradual deterioration of the environment in the Sundarban. Sea level rise, severe cyclonic storms, salinity, rainfall and erosion are very common threats to the fragile ecosystem in the Sundarban. Some mangrove species are on the way to extinct from the Sundarban. Fishes and other aquatic resources are severely affected by the changing water properties. Rice cultivation is widely hindered by the adverse climatic phenomenon. Now, the livelihood of islanders is in front of daunting challenges.

References

B. A. E. & S. (2011). *District statistical handbook, South 24 Parganas and North 24 Parganas*. Department of Statistics and Programme Implementation, The Government of West Bengal. Kolkata, New Secretariat Buildings: Bureau of Applied Economics and Statistics (B.A.E. & S). http://www.wbpspm.gov.in/publications/District%20Statistical%20Handbook. Accessed 22 Nov 2018.

Bhattacharya, S. (2007). *Lesson learn for vulnerability and adaptation assessment from India's first national communication*. Paper 7 of BASIC project, Winrock International. http://www.basic-project.net/data/final/Paper07India%20Lessons%20for%20VA%20from%20First%20National%20Communicat%85.pdf. Accessed 30 Sept 2016.

Bindoff, N., Willebrand, J., Artale, V., Cazenave, A., Gregory, J., Gulev, S., Hanawa, K., Quéré, L., & Co-authors. (2007). Observations: Oceanic climate change and sea level. In S. Solomon, D. Qin, M. Manning, Z. Chen, M. Marquis, K. B. Averyt, M. Tignor, & H. L. Miller (Eds.), *Climate change 2007: The physical science basis. Contribution of working group I to the fourth assessment report of the intergovernmental panel on climate change* (pp. 385–432). Cambridge: Cambridge University Press.

Chand, B. K., Trivedi, R. K., Dubey, S. K., & Beg, M. M. (2012). *Aquaculture in changing climate of Sundarban: Survey report on climate change vulnerabilities, aquaculture practices and coping measures in Sagar and Basanti blocks of Indian Sundarban*. Kolkata: West Bengal University of Animal and Fishery Sciences.

Dinse, K. (2012). *Climate variability and climate change: What is the difference?* http://www.miseagrant.umich.edu/wp-content/blogs.dir/1/files/2012/05/11-703-Climate-Variability-and-Climate-Change-2013update.pdf. Accessed 15 Feb 2013.

Doney, N. M., & Levine, C. D. (2006). How long can the ocean slow global warming? How much excess carbon dioxide can the ocean hold and how will it affect marine life? *Oceanus Magazine*. https://www.whoi.edu/oceanus/feature/scott-doney-and-naomi-levine. Accessed 8 Jan 2014.

DPD. (2009). *Human development report, south 24 Parganas*. Development and Planning Department (DPD), Government of West Bengal, Kolkata: HDRCC.

Ghosh, A. (2012). *Living with changing climate: Impact, vulnerability and adoption challenges in Indian Sundarbans*. New Delhi: Centre for Science and Environment (CSE). http://cseindia.org/userfiles/Living%20with%20changing%20climate%20report%20low%20res.pdf. Accessed 20 Dec 2013.

Goshthakur, D. (2009, December 4). Sundarbaner Bastutantra Nonarbish (Poison of Salinity in the Ecosystem of Sundarban). *Annadabazar Patrika*, p. 4.

Hazra, S., Ghosh, T., DasGupta, R., & Sen, G. (2002). Sea level and associated changes in the Sundarbans. *Science and Culture, 68*(9–12), 309–321.

Hazra, S., Samanta, K., Mukhopadhyay, A., & Akhandet, A. (2010). *Temporal change detection (2001–2008) study of Sundarban*. Final report. School of Oceanographic Studies, Jadavpur University. http://www.iczmpwb.org/main/pdf/ebooks/WWF_FinalReportPDF.pdf. Accessed 23 Jan 2013.

IPCC (2007a). *Impacts, adaptation and vulnerability*. Contribution of working group II to the fourth assessment report of the intergovernmental panel on climate change. Cambridge: Cambridge University Press.

IPCC (2007b). *Climate change 2007: The physical science basis*. Contribution of working group I to the fourth assessment report of the intergovernmental panel on climate change. Cambridge: Cambridge University Press.

Jhala, Y. V., Qureshi, Q., Gopal, R. (Eds.) (2015). *The Status of Tigers in India, 2014*. National Tiger Conservation Authority, New Delhi and the Wildlife Institute of India, Dehradun. New Delhi: NTCA

Khan, T. M. A., Singh, O., & Rahaman, M. S. (2000). Recent sea level and sea surface temperature trends along the Bangladesh coast in relation to the frequency of intense cyclone. *Marine Geodesy, 23*(2), 23–103.

Lal, M., Singh, K. K., Rathore, L. S., Srinivasan, G., & Saseendran, S. A. (1998). Vulnerability of rice and wheat yields in NW India to future changes in climate. *Agricultural and Forest Meteorology, 89*(2), 101–114.

Mistri, A. (2011). *প্রকৃতিক দুর্যোগ ও পরিব্রাজনঃ আয়লার পরিপ্রেক্ষিতে* (Natural hazard and migration: In the context of Aila). *Antorjatik Pathsala, 1*(2), 124–140.

Mistri, A., & Das, B. (2015). Environmental legislations and livelihood conflicts of fishermen in Sundarban, India. *Asian Profile, 44*(3), 389–400.

Mitra, A., Gangopadhyay, A., Dube, A., Schmidt, A. C. K., & Banerjee, K. (2009). Observed changes in water mass properties in the Indian Sundarbans (Northwestern Bay of Bengal) during 1980–2007. *Current Science, 97*(10), 1445–1452.

Mitra, A., Chowdhury, R., Sengupta, K., & Banerjee, K. (2010). Impact of salinity on mangroves of Indian Sundarbans. *Journal of Coastal Environment, 1*(1), 71–82.

Nandy, S., & Bandyopadhyay, S. (2011). Trend of sea level change in the Hugli Estuary, India. *Indian Journal of Geo-Marine Science, 40*(6), 802–812.

Rudra, K. (2010). *The proposal of strengthening embankment in Sundarban: Myth and reality*. Discussion Paper-I. http://www.mcrg.ac.in/rw%20files/RW35/7.Discussion_Paper1.pdf. Accessed 6 Jan 2014.

Rupa Kumar, K., Sahai, A. K., Kumar, K. K., Patwardhan, S. K., Mishra, P. K., Revadekar, J. V., Kamala, K., & Pant, G. B. (2006). High-resolution climate change scenarios for India for the 21st century. *Current Science, 90*(3), 334–345.

Singh, O. P. (2002). Spatial variation of sea level trend along the Bangladesh coast. *Marine Geodesy, 25*(3), 2005–2012.

Singh, O. P. (2007). Long-term trends in the frequency of severe cyclones of bay of Bengal: Observations and simulations. *Mausam, 58*(1), 59–66.

Singh, O. P., Khan, T. M. A., & Md Rahman, S. (2001). Has the frequency of intense tropical cyclones increased in the North Indian Ocean? *Current Science, 80*(4), 575–580.

STR. (2015). *Administrative units, Government of West Bengal*. http://sundarbantigerreserve.org/urls/administrative-unit.html. Accessed 19 June 2015.

Unnikrishnan, A. S., Kumar, K. R., Fernandes, S. E., Michael, G. S., & Patwardhan, S. K. (2006). Sea level changes along the Indian coast: Observations and projections. *Current Science, 90*(3), 362–368.

Occupational Issues in Sundarban 3

Abstract
This chapter deals with occupational issues faced by the inhabitants of Sundarban. They depend on farming and fishing, which are natural resource-based activities that are highly susceptible to climate stimuli. Farming is affected by seasonality, irrigational insufficiency, and other infrastructural issues. Fishing is constrained by environmental activism, which generally fails to consider the economic entitlements of fishers. Ecotourism falls flat because of the lack of participation by most local people; most of the revenue is drained by multinational companies and other stakeholders. The people of Sundarban, therefore, experience extreme marginality in their respective activities; they may not even get a job for more than 6 months at the local level. Thus, migration is often the best option to cope with the situation.

Keywords
Framing · Fishing · Ecotourism · Environmental legislation · Livelihood conflict

Farming and fishing are the principal activities of the Sundarban people. To date, agriculture in Sundarban has been technologically primitive or traditional agriculture. The farmers plough small areas of land with a yoke of cattle (cow or buffalo) and use a very minimal amount of fertilizers and pesticides. The land is farmed to provide sustenance. Fishing is also practiced for survival. The residents capture fish from rivers and creeks with human-driven wooden craft and tradition contrivances. Sustainability in the means of living is a very essential part of a livelihood. In Sundarban, both activities are facing different sorts of challenges, which are the focus of this chapter. This chapter also explores alternative means of living in Sundarban. The major data sources used for this discussion are the Census of India (1991, 2001, and 2011), Agricultural Census of India (1995–96, 2000–01, 2005–06, 2010–11, and 2015–16), and District Hand Books of South and North

A. Mistri, B. Das, *Environmental Change, Livelihood Issues and Migration*,
https://doi.org/10.1007/978-981-13-8735-7_3

24 Parganas (2004–2014) by the Bureau of Applied Economics and Statistics (B.A.E. & S.), Department of Statistics and Programme Implementation, Government of West Bengal.

3.1 Worker and Work Status

According to the 2011 Census, the total number of workers in Sundarban is 1.7 million and the work participation rate (WPR) is 38%, whereas the national and state (WB) averages for WPR are 39.8 and 38.1%, respectively. Approximately 59% of workers are engaged in farming (Census 2011) and 33% are engaged in fishing (B.A.E. & S 2013) in Sundarban. Intercensal comparisons (Fig. 3.1) reveal that the proportion of cultivators[1] declined to around 17% in 2011 from 24% in 2001. Agricultural laborers[2] increased to nearly 42% from 36% during that period. Workers in household industries slightly increased from 5% to 7% during 2001 to 2011.

Within the total population in 2011, around 22% were main workers who worked 6 or more months in a year and 15.2% were marginal workers who worked less than 6 months in a year (Fig. 3.2). The share of main workers has been declining gradually: from 27% in 1991, 24% in 2001, to 22% in 2011. However, the proportion of marginal workers in Sundarban have been increasing at a rapid pace: from 3% in 1991, 11% in 2001, to 15% in 2011. Marginal workers have increased threefold from 1991 to 2011, and they are highly concentrated in the agricultural labor force, accounting for 51% and 55% in the 2001 and 2011 Census, respectively. Out of the total workforce (or working population), main workers have declined to 59.4% in

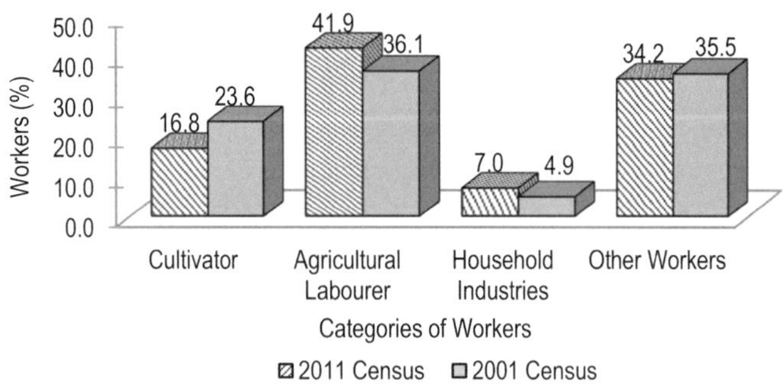

Fig. 3.1 Classification of workers in Sundarban, India

[1]A person is classified as a cultivator if he or she is engaged in the cultivation of land owned or held by the government, private persons, or institutions for payment in money, kind, or share (Census 2011).

[2]A person is defined as an agricultural laborer if he or she works on another person's land for wages in money, kind, or share (Census 2011).

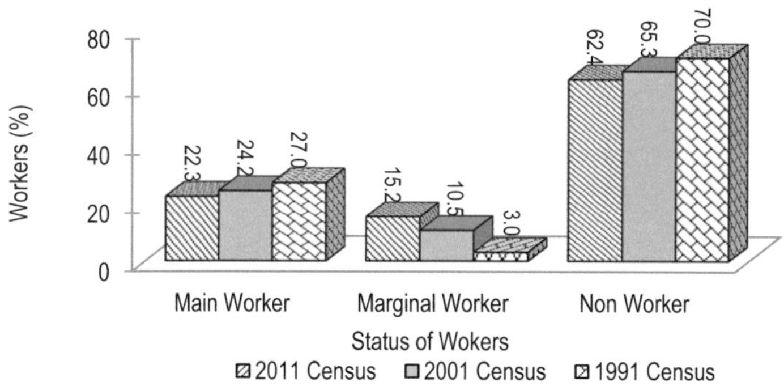

Fig. 3.2 Duration of work engagement for workers in Sundarban, India

2011 from 69.7% in 2001, whereas the share of marginal workforce has increased to 40.6% from 30.3% during that period.

Land holding in Sundarban is mostly marginal in size class[3] (63%) distibuted among the 86% owners. The increasing population in Sundarban—from 3.8 million in 2001 to 4.43 million in 2011, at a growth rate 1.65%/year (geometric growth)—has exerted tremendous pressure on cultivable land, resulting in conversion into settlements, vegetation, aquaculture, and the brick kiln industry (Hazra et al. 2010). Moreover, every year, a certain amount of cultivable land is lost to the river due to erosion or subsidence of the same (Ghosh 2012). Nearly 62% of total cultivable land in Sundarban is low lying and suffers from increasing salinity during the dry season, from intrusion of saline water and water surge, capillary action, and increasingly sulphate build up (DoSA 2018). Hence, farmers have to keep their land as a current fallow from time to time, resulting in a reduction of net area sown[4] (Fig. 3.3). However, declining cultivators, increasing agricultural laborers, and higher marginality (working less than 6 months per year) are a cause for concern in the livelihoods of the Sundarban people.

3.2 Agriculture and Its Issues

Agriculture in Sundarban is very seasonal and depends on the monsoons. Seasonality in agriculture leads to a long lean period that pushes the workers into marginal work status. Due to a lack of irrigation facilities, most of the farmers practice monocropping, especially *Aman* rice. Apart from rice, less water-intensive crops (e.g., barley, wheat, potato, musur, maskalai, khesari, til, mustard, linseed, and gram)

[3]Agricultural Census of India (2010–11) classifies the size of landholding as follows: marginal (below 1.0 hect.), small (1.0–1.99 hect.), semi-medium (2.0–3.99 hect.), medium (4.0–9.99 hect.), and large (10 hect. and above) size holdings.

[4]Net area sown represents the area sown with crops for at least one in any of the crop season of the year, counting areas sown more than once in the same year only once.

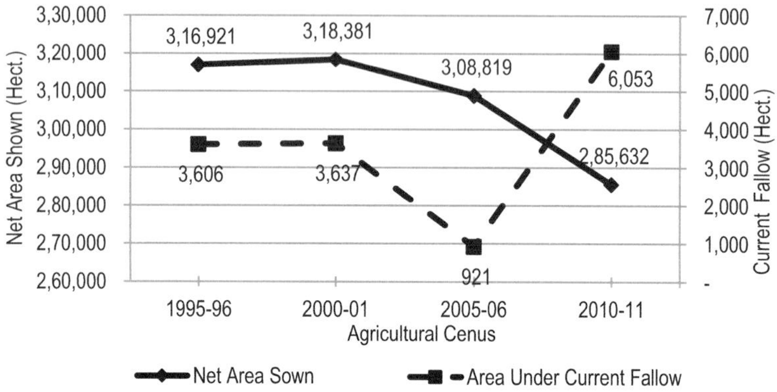

Fig. 3.3 Net area sown and area under current fallow in Sundarban, India

are cultivated in different parts of Sundarban in *Rabi* cropping. In the monsoon, low-lying agricultural land is filled up with water and naturally prepared for rice cultivation. The very few farmers who have sources of irrigation can practice double cropping or multiple cropping. The islands of Sundarban are surrounded by rivers that contain salty water. Inland tanks (50.8%) and canals (26.1%) that store fresh water during the rainy season are the major sources of irrigation in *Rabi* cropping (Agricultural Census, 2005–06). A vast area of agricultural land in Sundarban does not have proper irrigation facilities. More than half (57.3%) of the net area sown remains wholly un-irrigated, 29.7% is partially irrigated, and only 13.0% is wholly irrigated (Agricultural Census, 2010–11).

Available official data on the last 10 years of rice production from the Bureau of Applied Economics and Statistics (B.A.E. & S), Government of West Bengal shows that the yield rate (kg./hect.) of *Aman* rice (Plates 3.1 and 3.2) remained below the level of 2004–05 until 2010–11; after that, it followed an increasing trend (Fig. 3.4). The yield rate in Sundarban (2391 kg./hect. in 2012–13) is far below the state average (2596 kg./hect.). Although the proper data set is not available, there was a complete crop failure in 2009–10 due to inundation by 8 to 10-meter-high surge due to the severe cyclonic storm Aila on 25 May 2009. Even after 2–3 years, it caused low yield rates. Figure 3.3 shows that the area under current fallow has increased greatly: from 921 hectares in 2005–06 to 6053 hectares in 2010–11, an increase of more than 6.5-fold. As a result, the net area sown has sharply dropped to 285,632 hectares from 308,813 hectares during that period.

Apart from climatic vagaries, poor agricultural infrastructure affects the production and reduces the yield rate. Water outlet systems and fragile mud build embankments are main concern of the farmers. Low-lying islands in Sundarban are protected by a 3500-km-long mud build embankment. The rise of high tide varies by 2–5 meters in Sundarban. Hence, outletting the rain water is not always possible naturally; farmers must wait for the low tide, until it reaches below the average

Plate 3.1 A farmer harvesting rice with a sickle

Plate 3.2 Rice threshing by a locally innovated technique, revolving the threshing machine by an engine attached with a belt of rope

(absolute) height of the island. A long submergence of paddy saplings may rot the stems, which leads to a low yield rate of rice.

Due to the lack of maintenance and community-based management of freshwater reservoirs, the areas under irrigation by these tanks and canals have been gradually decreasing over the census years. On the contrary, the areas under irrigation by tubewells, especially shallow boreholes, have emerged as a major source of irrigation (Fig. 3.5). However, irrigation by tubewells is not a reliable source of irrigation

Inland tanks (50.8)

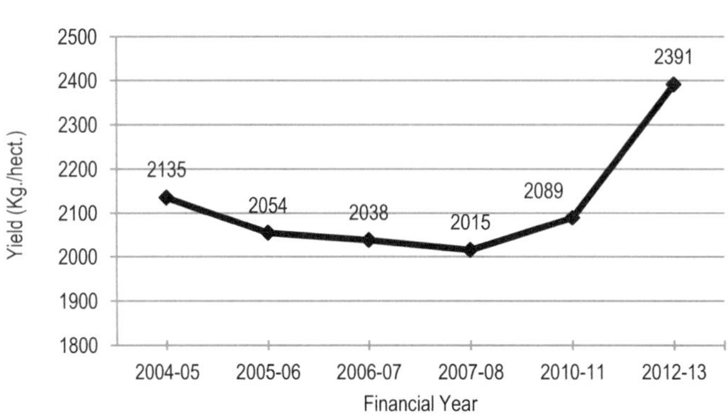

Fig. 3.4 Yield rate (kg./hect.) of Aman rice production in Sundarban, India

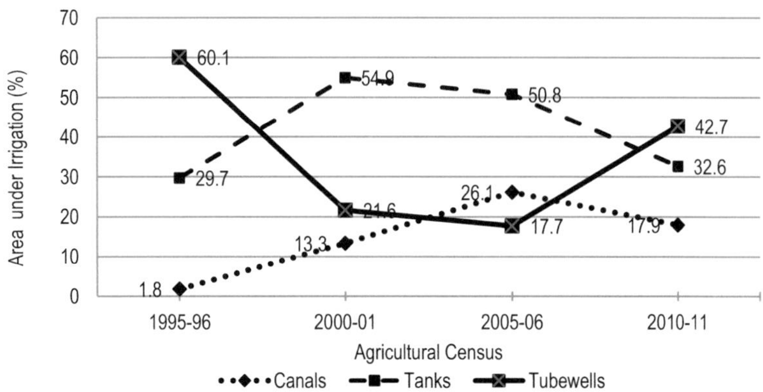

Fig. 3.5 Net irrigated area under different sources of irrigation in Sundarban, India

in Sundarban in the long term. In 1995–96, its share was 60.1%; after that, it has declined to 21.6% and 17.7% in 2000–01 and 2005–06, respectively, and again increased to 42.7% in 2010–11.

The use of ground water is not a sustainable use of the resource in Sundarban, as it causes environmental imbalances. Excessive extraction of ground water leads to the intrusion of saline water, which causes salinization of the top soil and ultimately hampers the production. Sundarban is a severely arsenic (more than 300 μg/L) affected area or red-alerted zone (SOES 2010). The extraction of ground water snowballs the health hazards. In addition, credits or subsidies from cooperative societies and Gramin Banks, storage facilities, use of fertilizer, hybrid seeds, modern techniques, and electricity facilities are in awful condition (Mistri 2012; B.A.E. & S 2011).

3.3 Fishing and Its Production

Apart from farming, fishing is another principal means of livelihood in Sundarban. Around 0.57 million people—that is, 12.4% of the total population or 33% of the total workers—were engaged in fishing in 2010–11 (B.A.E. & S 2013). Although the fishing activities have increased from 19% of the total workforce in 2000–01 to 33.0% in 2010–11, tremendous ups and downs have been experienced (Fig. 3.6). In 2005–06, the greatest number of people were engaged in fishing—that is, nearly 1.5 million. However, this number sharply declined to 0.36 million in 2007–08; after that, it again registered a rise and fall at a very slow pace in 2010–11 and 2012–13, respectively.

The production of fish has also varied over the financial years. Since 2007–08, production has shown a declining trend. Until the mid-2000s, tiger shrimp were abundantly available in rivers and creeks; they were easily caught by a drawing net with the finest weave (Plates 3.3, 3.4, 3.5 and 3.6). Nearly four lakhs (0.4 million) people, especially women and children, were involved, with approximately 1500–-3000 million shrimp caught during 1995–2005 (Feature Box 3.1). The collection of tiger shrimp was a huge economic gain that boosted the per capita income in Sundarban. A 1-inch-long tiger shrimp sold for Rs. 2–5 each in 2005. However, the rampant harvesting severely destroyed the riverine ecosystem, increasing river bank erosion, destroying mangroves, and killing immature carps and eggs. Therefore, it was a short-lived activity, and the production of tiger shrimp abruptly declined within the next 5 years. Today, some people are engaged in this activity, but not on a wide scale.

In Sundarban, both capture and culture fisheries (aquaculture) are practiced. Low-lying areas and the estuary locations of the islands in Sundarban are suitable

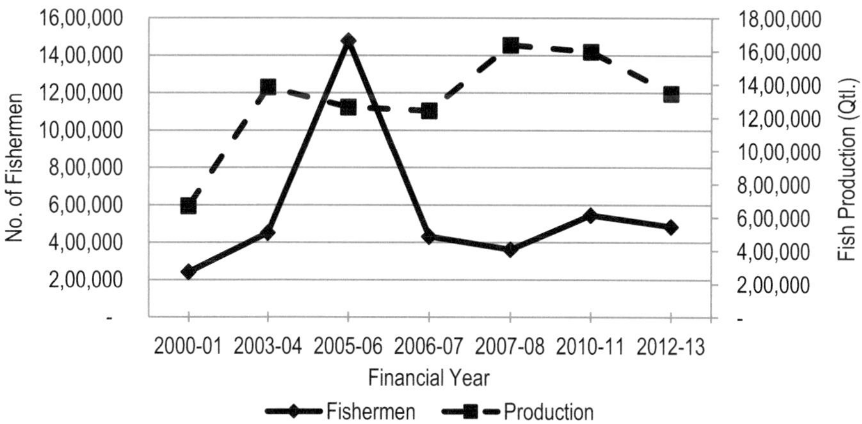

Fig. 3.6 The number of persons engaged in fishing and its production in Sundarban, India

Plate 3.3 A group of women and children collecting tiger shrimp

Plate 3.4 A local boy collecting tiger shrimp by drawing of a net

for brackish water aquaculture. Tiger shrimp and some economically valuable fish, such as *Parshe*, *Bhetki,* and *Pabda,* are widely cultivated. However, ownership is highly controlled by corporations and wealthy individuals. They accumulate the land through lease agreements at Rs 10,000–12,000 per annum and set up a farm where employment of local fishermen is limited.

Plate 3.5 A women drawing a net against a strong tidal force along the river bank

Plate 3.6 A women counting thin shrimp with a spoon-like contrivance and putting them into a dish

Feature Box 3.1: Tiger Shrimp and its Collection

The tiger shrimp is a special species of shrimp (Plate 3.7). Its yellow head and tail, along with black stripes along the body, give them a distinct appearance. Shrimp collection was a booming economic activity during the mid-1990s to the mid-2000s in Sundarban. One shrimp carp is 1–1.5″ long. It is collected by drawing a net with very tiny mesh along the riverbank during the low tide period against the tidal current (Plates 3.3, 3.4, 3.5 and 3.6) or by drawing a bag net from an anchored boat (Plate 3.8). In 2005, tiger shrimp were sold at Rs. 2–5 per carp, and a minimum income of Rs. 500–600 per day with the drawing net was possible with not-so-tedious work. A middleman or merchant collected the carp from the harvesters and resold them to fisheries. Next, the

(continued)

Feature Box 3.1 (continued)
carp were nursed in the brackish water of the fisheries until becoming 15- to 20-cm long; then, they were collected and sold at market or exported. The tiger shrimp has huge economic value. Today, the market price of 1.0 kg. of tiger shrimp is Rs. 1000–1200.

More than four lakhs (0.4 million) people, especially women and children, were engaged in the collection of tiger shrimp until the mid-2000s. The shrimp collection severely hampered the environment due to massive river bank erosion, the destruction of mangroves, killing of immature fish carp and crab, as well as other critically endangered species like tortoises and turtles. Meanwhile, it is associated with severe health hazards such as skin cancer, other diseases, and ill health. Contact with saline water makes women highly susceptible to vaginal cancer (Even-Paz and Shani 1989). Apart from health hazards, other socioeconomic issues, such as a high rate of school drop-out, illiteracy, and inequality, are highly associated with this activity.

Plate 3.7 Tiger Shrimp

Plate 3.8 Riverine fishing with traditional craft

Plate 3.9 Trawlers docked in Fraserganj fishing harbor in South 24 Parganas

A substantial proportion of fishers depends on capture fishing. They harvest wild fish and crustaceans from rivers and estuaries with the help of non-motorized boats (Plate 3.8), different kinds of nets, and other traditional contrivances. Marine fisheries (in the Bay of Bengal) are also practiced here with the help of trawlers and modern fishing equipment, but from only one fishing harbor: Fraserganj in Namkhana (Plate 3.9). These practices are very limited and are dominated by wealthy individual or corporate ownership. Local fishermen work as daily wage laborers. Since 2007, a restriction has been imposed on capture fishing: fish can only be caught for 9 months, August to March; fishing during the other 3 months, April to July, is strictly prohibited because it is the breeding period of fish. Traditional fishermen are socioeconomically weaker in Sundarban. They generally reside along the edge of the forest or along the river in shanty houses upon the mud-build embankment. Most of them have no access to electricity, safe drinking water, or clean sanitation facilities. They have minimal or no availability of basic amenities and assets. Fishing is their only means of sustenance. The three-month off-session has pushed them into abject poverty and forced them to move out of Sundarban to find a job.

3.4 Non-timber Forest Products

Fishing in Sundarban also includes the collection of non-timber forest products (NTFPs), such as honey, beeswax, firewood, poles, tannin bark, and nypafruticans (*golpata*). Environmental legislation for the conservation of the reserve forest immensely affect the gathering of NTFPs. The collection of timber, firewood, and poles has not been permitted since the Amendments of the Wildlife Protection Act (WLPA) in 2006. The production of timber, firewood, and pole in the district of

South 24 Parganas has shown a declining trend from 2006–07 onward, although regular data for all financial years, particularly for the Sundarban Reserve Forest, are not available (Table 3.1).

More than 75% of the area of the Sundarban Biosphere Reserve (SBR) is under South 24 Parganas; hence, statistics related to the district provide a fair estimation. The WLPA Amendments of 2006 has also been imposed to restrict the illegal collection of honey and wax without permission and channelized the collection through the legal process. Therefore, honey and wax production have increased during that period, but it lacks consistency. Both the expenditure and revenue

Table 3.1 Forest production, revenue, and expenditure in South 24 Parganas

Year	Timber (1000s of m^3)/ Quintal	Fuel (1000s of m^3)/ Quintal	Pole (Number)	Honey (Quintal)	Wax	Revenue (1000s of Rs)	Expenditure
1999–00	700 ('000 Qtl)	13,980 ('000 Qtl)	–	–	–	2890	36,072
2000–01	–	–	–	–	–	1342	47,246
2001–02	–	–	–	–	–	1303	55,393
2002–03	0.051 ('000 m^3)	0.16 ('000 m^3)	359	–	–	3518	62,073
2003–04	0.06 (")	0.073 (")	1097	–	–	9475	58,048
2004–05	0.054 (")	0.921 (")	414	–	–	11,325	57,965
2005–06	0.165 (")	0.482 (")	64	–	–	10,638	97,479
2006–07	7.9 (")	355.68 (Qtl.)	5735	130.37	2.60	11,118	136,721
2007–08	5.01 (")	309.50 (Qtl.)	119	159.74	4.23	12,519	155,935
2008–09	1.403 (")	352.30 (Qtl.)	466	132.70	8.89	13,216	196,676
2009–10	0.5 (")	115 ('000 m^3)	138	119.50	0.40	9446	271,028
2010–11	–	–	–	200.75	18.77	13,093	298,493
2011–12	2.06	250.62 (Qtl.)	2482	9915.25	18.43	7317	143,657
2012–13	..	..	..	8867.05	..	47,442	45,336
2013–14	..	..	..	392.00	..	16,778	214,104

Source: B.A.E. & S, District Statistical Hand Book, 2004 to 2013

collected by the Forest Department in South 24 Parganas have increased during 1999–00 to 2013–14. The expenditure has surpassed the revenue to a great extent. This growth has occurred since 2006 onward.

3.5 Conservation and Livelihood Conflicts

Forest conservation and livelihood conflicts of "ecosystem people"[5] in different parts of India have become widely debated and common political issues in recent times. Sustainable development is a totality of environmental, economic, and social sustainability. However, in India, forest conservation is highly skewed toward environmental sustainability and hardly balanced with its counterparts. It immensely affects the ecosystem people by forcing them to move away from their usual activities, resulting in agitation, strike, and migration in different parts of India (Kothari and Patel 2006). Some highly controversial issues include political unrest over the Western Ghats in Kerala, the Goa mining suspension, the eviction of residential villages from the Sariska Tiger Reserve in Rajasthan, and the expulsion of the Gujjars nomadic pastoralists in the Rajaji National Park in the Himalayan foothills.

The present study area of Sundarban clearly illustrates such conflict. Since the formation of the Sundarban Tiger Reserve (STR) in 1973 within the Sundarban Reserve Forest (SRF) under the central government's flagship program ("Project Tiger") per the WLPA, the livelihood of traditional fishers has been jeopardized. The gradual implementation of new regulations under the WLPA of 1972 and its 2006 Amendment for better conservation of the reserve forest has hindered fishing activity. Conservation efforts to reduce the effects of global warming and set up a suitable habitat for endangered species, such as royal Bengal tigers, Irrawaddy and Ganges River dolphins, and estuarine crocodiles, are widely appreciated. However, they conflict with the means of livelihood of the socioeconomically depressed people of the islands and curtails their economic freedom to utilize the water and forest resources for the purpose of consumption or exchange. Fishing in Sundarban has a long history and is a generational occupation. The fishers have engaged in forest-based economic activities since civilization began in Sundarban. The intense conservation interest of organizations such as UNESCO and WWF suppress the economic entitlement of the fishers, which is a violation of human rights (UNO 2015)—"the right to work, to free choice of employment, to just and favourable conditions of work and to protection against unemployment" (Article 23, Universal Declaration of Human Rights [UDHR]) and "the right to an adequate standard of living" (Article 25, UDHR).

During the formation of the STR, its "Management Plan" mentioned that only non-motorized boats would be allowed in the buffer areas for fishing. Motorized

[5]The people, especially communities, who depend on the natural environment of their locality to meet most of their material needs are known as "ecosystem people" (Kothari and Patel 2006).

boats are strictly prohibited. In the buffer that surrounds the core, access is allowed for livelihood purposes with a proper license and permit. Human interference is strictly prohibited in the core areas. Details on the restrictions and their associated conflicts are discussed in the following section.

3.5.1 Limited Number of Boat License Certificates

The boat license certificate (BLC) is a kind of gate-pass for man-driven or non-motorized boats to enter into the buffer areas of Sundarban for fishing. It was introduced in 1980. Only 923 BLCs were issued for the STR at that time, and the number of BLCs has remained the same to the present date. Over the time, the number of fishers have increased, but not a single addition BLC has been issued; in fact, the number of BLCs decreased to 709 in 2009 (DISHA 2009). A huge proportion of fishers, therefore, are compelled to catch fish in the buffer area without BLCs, resulting in an increasing number of offenses and fines (DISHA 2006).

3.5.2 Contravention of the Act to BLC Distribution

There was a serious violation of Sect. 21 of the 1972 WLPA at the time of BLC issuing. Per the regulation, to set up a tiger reserve, a notice should be circulated in the regional language in the nearby towns and villages to inform the fishers or others whose livelihoods depend on the area of demarcation to register their boats or other contrivances used for the activities therein. The time duration for receiving applications for BLCs and other permits/licenses must be around 2 months. During the field investigation, the local fishers complained that the notice was posted only in the beat-offices in English for a month. Most of them who lived in remote areas were unaware of the notice, which resulted in a low number of applications being submitted; thus, few BLCs were issued. This was a serious violation of provisions of the regulation that meted out injustice to the fishers.

3.5.3 BLCs become Leasable Property

A limited number of BLCs were issued and were not properly distributed among the actual fishers who most needed them. BLCs are non-transferable except to blood-related kin and successors. After the death of a fisher, it is transferred to his sons or other relatives irrespective of their engagement in fishing. Local fishers complained that most of the BLC holders were comparatively wealthy persons with no association to fishing.

In an investigation of fishing households, it was observed that nearly 86% of the households did not possess any BLCs. Fishers borrow a BLC from the owner on a contract ranging from a month to a year. The rent for a BLC was approximately Rs. 15,000–20,000 per year in 2014–2015. Thus, BLCs have become a leasable property in Sundarban. Although BLCs are considered as a gate-pass for entry to catch fish in the Sundarban, there are lots of irregularities in the distribution of the limited number of licenses.

3.5.4 Legislation Supremacy over Humanity

In Sect. 38 V(4) of the WLPA Amendment Act of 2006, it is clearly mentioned that whenever a State Government prepares a tiger conservation plan, the government must ensure the agriculture, livelihoods, development, and other interests of the people living in the tiger-bearing forest or a reserve. In the British Period (1928), with the establishment of the SRF according to the Indian Forest Act of 1927, inhabited and forest areas were separated by the rivers (Table 3.2). Since then, no settlements have been allowed in the reserve forest. In the Marichjhapi incident, Bangladeshi refugees from Marichjhapi Island (one of the forested islands in north-west Sundarban) were forcibly evicted in 1979. Although the islanders of Sundarban have not been living in the forest since 1928, they highly depend on the forest for their livelihood. However, the WLPA Amendment of 2006 does not recognize the interests of people residing outside of the conservation area. The STR authority, therefore, is not legally obligated to consider the livelihood interests of the fishers. For the same reason, the Forest Rights Act of 2006 does not protect the livelihoods of the Sundarban people. This act recognized only the rights and occupation in forest land of a forest-dwelling Scheduled Tribe and other traditional forest dwellers who

Table 3.2 Conservation history of Sundarban, India

Year	Rules and regulations or legal designations
1878	Declared as Protected Forest
1928	Declared as a Reserve Forest under the Indian Forest Act, 1927
1973	Declared as Sundarban Tiger Reserve under Project Tiger, 1973 and core area formation
	Motorized boats for fishing strictly prohibited
1976	Sajnekhali Wildlife Sanctuary was designated
1980	Boat License Certificates were introduced
	The permit procedure was introduced
1984	The core area of STR was declared as a National Park
	West Bengal Fisheries Act was launched
	Fishing of specified size, group, or species of fish
	Use of fixed engine and other contrivances
	Dimension of net, size of mesh, and other contrivances to catch particular species and their mode of use
1987	The National Park was inscribed as a Natural World Heritage Site by UNESCO
2001	The biosphere reserve was recognized by UNESCO under the Man and the Biosphere Programme (MAB)
2007	The seasonal pass was introduced
	The Sundarban Tiger Reserve was newly designed and enhanced (core and critical tiger habitat)

Source: Authors' collection from various official documents

had been residing there for at least three generations prior to 13 December 2005. This is a clear-cut contradiction between the "legitimacy of the demand of human rights" and human rights as social ethics (Sen 2016). If a claim has enough plausibility to receive universal consent or ethical value, it must be considered as a human right and the law should be directed to secure it. The livelihood interests of the fishers in Sundarban are not protected because the existing regulations do not sanction the demand. Therefore, it is not justified as a right of fishers.

3.5.5 Shrinking the Area under Fishing

The gradual decline of buffer areas where fishing is allowed provisionally is one of the great constraints of fishing. The total area of STR is 2584.89 km^2, which is approximately 61% of the total SRF (4263 km^2). In 2009, through the newly designed Critical Tiger Habitat (1699.62 km^2), the core area increased to nearly 66% from 52% in 1984, thus shrinking the buffer areas to 34% from 48%.

3.5.6 Seasonal Pass–related Issues

Every fisher is required to issue a seasonal pass at the rate of Rs. 40 per annum (in 2014–15) to catch fish with a certified boat (BLC holding boat). Since 2007, the seasonal pass has been issued only for 9 months, from August to March, instead of the whole year. Three months, April to June, have been declared as the "closed season" for fishing because it is the breeding period for fish. However, it is unclear how the fishers can maintain their livelihood during these 3 months. The answer is not yet known by the STR authority.

3.5.7 Permit related Issues

Apart from BLCs and seasonal passes, fishers also need a permit to catch and trade fish. This permit was introduced at the time of the BLC launch in 1980. The permit carries information on the name of the crew, their life insurance policy numbers, and a description of fishing contrivances, which includes nets, cages, fishing fence, traps, and a fixed engine to make the boat stationary. A maximum of four people, including the BLC owner, are permitted to fish in a boat. The permit is issued in the name of the BLC owner at the rate of Rs. 5.0 per person per week for 42 days. After 42 days, it must be renewed; otherwise, a fine is levied at a certain rate for overstaying in the fishing ground. Often, after 42 continuous days of fishing, the fishers are too exhausted or sick, not willing to leave immediately to catch. They then delay renewal of the permit. However, the forest authority does not entertain excuses and levies the fine. Moreover, to renew a permit, a certain number of days are required. There is no provision to pay the man-days to the poor fishers and those days are wasted.

3.5.8 Restriction on Specific Species to Catch

The West Bengal Inland Fisheries Act of 1984 and its Amendments in 1993 and 1997 imposed several restrictions on fishing activity, including specification of size, group, or species and the use of fixing gear for catching a particular group of fish. The latter restriction is clearly mentioned on the permit; violations result in penalties, cancellation of permits, and even seizure of the contrivances. The fishers of Sundarban are economically very weak. One set of nets or contrivances is their sole property. They also borrow from others at a certain rate per annum or per fortnight basis. Without any assistance from the government, it is extremely difficult for the fishers to procure different kinds of nets or contrivances to catch various species because of the expense.

3.5.9 Penalties and Offenses

The forest department has categorised offenses into three groups (Table 3.3): prosecution report or charge sheet cases (POR), compounded offense report (COR), and offenses detected but offenders not found (UDOR) (DISHA 2009).

In COR, offenders are fined but not charge-sheeted and sent to court for trial. The COR have increased sixfold from 2000–01 to 2007–08 (DISHA 2009). The fishers of Sundarban are marginalized people belonging to Scheduled Caste (SC) and Scheduled Tribe (ST) communities. Most of them are illiterate. They are unaware of the existing restrictions, newly implemented and designated territories, categories of offenses, and penalties scheduled for different kinds of offenses. The violations and other details are printed on the back side of the BLC in English, which is not comprehensible by the fishers. Whenever they are arrested by the mobile forest guards and fined, they think that forest personnel are intentionally harassing them. Sometimes, forest guards also take advantages of their ignorance. Local fishers reported that forest guards forcefully take away the best-quality catch from boats; being unwilling to bend down to their demands results in abusive

Table 3.3 Types of offenses identified in SBR, India

Offence	List of offences
Prosecution report or charge sheet cases (POR)	Hunting of wild animals and birds, cutting of timbers and logs, destruction of the forest, kidnapping and robbing fishers, etc.
Compounded offense report (COR)	Fishing without BLC, permit, and seasonal pass or without renewal of these; more fishermen than actually permitted in a boat; overstaying (more than 42 days) and illegal entry to catch more in core areas; catching of restricted species; collecting of wild fruits and firewood, etc.
Offenses detected but offenders not found (UDOR)	Both the POR and COR

Source: Developed on the basis of DISHA 2009

language, false convictions, fines, and even physical abuse such as smacking, kicking, and being hit with the gunstock. No program has been conducted yet to increase the awareness of the fishers regarding the kinds of offenses, fines scheduled for different offenses, the area of restrictions, and so on. Therefore, they are not empowered to raise their voices for their rights. Conflicts between humans and tigers is a regular occurrence in Sundarban. There is lack of universal coverage of accidental compensation. A tiger-victim family is compensated Rs. 100,000; however, this money is only provided to legal fishers. Every year, many fishers who become victims of tigers are not reported properly.

Ultimately, the economic entitlement of the poor fishers has always been overwhelmed by environmental activism. To conserve a protected area, rigorous implementation of environmental legislation is not enough; local people's participation at large is also essential. That was the prime aim in the formation of a biosphere reserve (9630 km^2) by UNESCO in 2001, which involved local people in the conservation process and its sustainable use. However, the authorities have been undermining the regulations that secure the rights of fishers and forest dwellers in order to fulfill their commitment to the international community. As a result, a group of fishers have been compelled to divert from fishing. It begs the question: whose Sundarban is it—the islanders' or the tigers'?

3.6 Other Activities

Apart from farming and fishing, the people of Sundarban are engaged in small business, such as shopkeepers, hawkers in local markets or tourist spots, and sellers of poultry, fish, vegetables, green coconuts, betel, and betel nut in local markets. Men are also engaged in tailoring, carpentry for making boats or crafts, and making date palm jaggery (*nolen-gur)* (Plate 3.10). Women are primarily engaged as *beedi laborers*[6] (Plate 3.11), rice wine makers (*Hariya*),[7] especially among the tribes, and makers of handicrafts with bamboo, coconut leaves, palms (Plate 3.12), and embroidery work on clothes. A small proportion of workers is engaged as daily or casual laborers in someone's home, or in digging and carrying soil for repairing embankments and ponds or road construction (Plates 3.14 and 3.15). Transportation and communication activities, such as paddle vans, engine vans,[8] and drivers of automobiles, boats, trawlers, and steamers account for a fair proportion of workers. Sundarban is still far removed from the din and bustle of the city life of Kolkata. Entertainment includes traditional folk and ballad singers and local drama groups. A

[6]*'Beedi'* is Indian cigarette filled with tobacco flake and wrapped in a leaf of the Kendu tree. Women prepare and sell them for Rs. 80–100 for every 1000 pieces.

[7]*'Hariya'* is a local liquor prepared from rice fermentation. It is widely famous among the tribes. One tribal family earns an average of Rs. 2500–3000/week selling it at weekly arranged local markets.

[8]An engine van is made by setting up an engine in a paddle van.

Plate 3.10 Preparation of date palm jaggery (*Nolen-Gur*): boiling the sap of date palm to make it dense and pouring it into a jar

Plate 3.11 A group of women preparing *'beedi'* by wrapping dry Kendu leaf and tobacco flake

group of people survive on this. They are in demand at all festivals and family functions. Tourist lodges and hotels also arrange traditional shows at the rate of around Rs. 1000 per night (in 2016) (Plates 3.16 and 3.17). Initially, these entertainers used to sing folk stories of local deities, such as Bonbibi, Shajangali, and Dakhin Rai (Feature Box 3.2) and the snake goddess (Mansa Devi). However, now they have introduced a touch of modernization and often sing scintillating Hindi movie songs.

Plate 3.12 Tribal women preparing wrappers by needle and cloths, and weaving cushions with the dry leaves of date palm

Plate 3.13 Fish are drying in the sun on the sea beach, Fraserganj, South 24 Parganas

Plate 3.14 A group of people digging and carrying soil to repair a river embankment

Plate 3.15 Woman repairing a surfaced road damaged by super cyclone, Aila in 2009

Plate 3.16 An idol of Bonbibi flanged with Dukhe (on the left), Shahjangali, and Dokkhin Rai (on the right)

Plate 3.17 A local drama group performing *Dukhe Yatra* (drama)

Feature Box 3.2: Deity Bonbibi—An Imprint of Cultural Harmony in Sundarban

A unique cultural harmony in Sundarban has been created and is centered on the common resource—the forest. Religion, caste, and class distinctions have been broken down. Hindu and Muslim people share a common deity, Bonbibi. Bonbibi is worshipped along with her brother, Shahjangali, and Dokkhin Rai. Bonbibi is the goddess of the forest. Bonbibi is worshipped mostly by the fishers, honey collectors, and the woodcutters before entering into the forest to protect them from tiger attacks. Bonbibi is believed to be a daughter of a fakir, named Ibrahim from Mecca. Ibrahim had two wives, Phulbibi and Golalbibi.

(continued)

Feature Box 3.2 (continued)

Meanwhile, Allah decided to send Bonbibi and Shahjangali from heaven for a divine mission. When Golalbibi became pregnant, Ibrahim left her in a forest to satisfy his first wife Phulbibi's wish. Golalbibi had given birth to twins, Bonbibi and Shahjangali. Golalbibi abandoned Bonbibi in the forest and left with Shahjangali in her arms. Bonbibi was raised in the forest by a doe. After 7 years, Ibrahim realized his mistake and returned Golalbibi and the two children to Mecca.

Dokkhin (meaning "south") Rai (meaning "lord/king"), the lord of the south, is known as the demon king. Actually, he was a Hindu sage living in the forest of Sundarban. He had supernatural powers to transfer himself into anything he wanted. The 370 million spirits, demons, goblins (*bhoots, prets, dakinis, deo*), and tigers are under his dominion (Jalais 2011). He gradually spread his kingdom all over the Sundarban. He stopped sharing the forest resources with human beings and killed them on the ground that it was a tax (*kawr*) for the products that they were collecting.

Once, while praying at the mosque, Bonbibi and Shahjangali received two magical hats. With the help of those magical hats, they flew to the country of eighteen tides (*atharo bhatir desh*) in Hindustan. After reaching there, Shahjangali practiced the *azan* (call to prayer). When the sound of *azan* reached the ears of the demon king Dokkhin Rai, he came to fight them, along with his mother, the goddess Narayani, and her army of ghosts and goblins. Bonbibi defeated Narayani and Dokkhin Rai after a long battle. Bonbibi and Sajanghali forgave them and returned the half of the erstwhile kingdom of Narayani and her son. Although the inhabited part of the Sundarban is believed to be the kingdom of Bonbibi, Dokkhin Rai is believed to be the ruler of the deep forest.

Story of Dukhe:

Once, there were two *Moule* (honey collector) brothers, Dhona (meaning "wealthy man") and Mona in a village named Barijhati. Dhona planned to go for a trip with a fleet of seven boats to collect honey in a *mahal* (dense forest) of the country of the eighteen tides. He took a poor village boy, Dukhe, along with him. Dukhe (meaning "sadness") was a son of a widow and belonged to a backward caste. Before leaving for the trip, Dukhe's mother advised him to remember Bonbibi in case he got into any serious trouble. The fleet reached the Kendokhali char, which was a part of the realm of Dokkhin Rai. Due to Dakkhin Rai's magical power, the group was not able to collect any honey and wax for 3 days. On the third night, Dakkhin Ray appeared in the dream of Dhona and asked for a human sacrifice as a *kawr* (tax). Greedy Dhona agreed to sacrifice Dukhe in exchange for honey and wax. So, after collecting enough wax and honey, Dhona left Dukhe in the forest and returned to the village. When Dukhe was about to be killed by Dokkhin Ray in the disguise of a tiger,

(continued)

Feature Box 3.2 (continued)
he started chanting a prayer invoking Bonbibi. On hearing his chant, Bonbibi came along with her brother Shahjangali. Shahjangali defeated Dokkhin Rai and saved Dukhe. Then, Bonbibi sent her pet crocodile, *Seko,* to drop him to his village with a generous amount of wax, honey, and other precious items. After returning to the village, the backward boy, Dukhe, popularized the worship of Bonbibi in the neighborhood. Thus, a Muslim girl is worshipped by her Hindu followers as *Bondurga* or *Bondevi* or Bonbibi.

3.7 Job Diversification and Prospects

Agriculture in Sundarban suffers from extreme seasonality, and fishing conflicts with environmental regulations. Moreover, both are natural resource-based activities that are highly susceptible to natural calamities. Job opportunities in daily or casual labor, household industries, transportation, and communication are very minimal and also follow the seasonality. Therefore, a substantial number of people, those who are staying in Sundarban, cannot get a job for more than 6 months in one singular occupation. They have to diversify their jobs at the local level. During the field investigation, it was observed that nearly 54% of non-migrants participated in job diversification during the previous 365 days. The change was observed more among workers who were engaged in agriculture (38%) and fishing (28%) activities. However, job diversification highly depends on the availability of employment opportunities. A lack of jobs leads to migration to cope with economic needs.

Ecotourism is the best alternative option to agriculture and fishing in Sundarban. When forest conservation efforts affect the livelihood opportunities of the fishers (Mistri and Das 2018), ecotourism balances between both. It promotes participatory conservation by encouraging community involvement and ensures their livelihoods. Moreover, for those who are unwilling to leave Sundarban for a job, ecotourism may be a good option. Ecotourism in Sundarban has enormous potential due to its worldwide reputation as a world heritage site, global biosphere reserve, and variety of natural attractions, such as magnificent wilderness of mangroves, ferocious Royal Bengal Tigers, estuarine crocodiles, river terrapins, and different species of deer, waterfowl, and migratory birds. Moreover, there is adequate infrastructure for lodging in hotels, guest houses, tourist lodges, and on boats or launches (Bhattacharya et al. 2011). Connections with the state capital Kolkata by train, rail, bus or private cars have also improved.

However, ecotourism has not been successful because of the local peoples' lack of interest. Their participation in the industry is very small in proportion (Bhattacharya et al. 2011). Coordination among local people and their large-scale involvement are the major criteria for a flourishing ecotourism industry.

In the study, out of 400 households, only 24 people from 22 households were engaged in ecotourism. They were mainly engaged in petty jobs, such as van and boat or launch drivers, cooks in hotels or lodges, artifact sellers, cameramen, folk dancers, sellers of green coconut and fast food (e.g., velpuri, golgappa, fish fry), and operators of temporary *Pan* (betel) and *Beedi* (cigarette) shops (Plate 3.18, 3.19, 3.20, 3.21, 3.22 and 3.23). Most of the hotels, tourist lodges, and guest houses are owned by either multinational companies or wealthier people from outside of Sundarban. They generally do not employ local workers. A tour package, where the tourist agencies arrange the entire safari and accommodations, is widely popular

Plate 3.18 Green coconuts being sold in front of a tourist lodge, Sajnekhali, South 24 Parganas

Plate 3.19 Papri-Chat (Velpuri) in the sea beach, Bakkhali, South 24 Parganas

Plate 3.20 A tourist enjoying a delicious sea fish fry, Bakkhali, South 24 Parganas

Plate 3.21 A photographer with camera; sitting on a chair at Rs. 5/hr; two women deal the business, standing at the farthest corner in the photo

in Sundarban. Local people or boatmen generally do not get a chance to enter into the business cycle. The whole tourism industry, therefore, is controlled by outsiders or stakeholders. A small proportion of the total revenue generated from ecotourism trickles down to the local people (Bhattacharya et al. 2011). This is the one of the main reasons for the reluctance of local people to get involved in tourism. Seasonality is another drawback of tourism in Sundarban. The peak season of tourism in Sundarban is approximately 4 months, from November to February. Summer and rainy seasons are avoided because of the extreme humidity, rain, and cyclones or storm surge. Therefore, people cannot choose tourism as a

Plate 3.22 A ferry crew serving between the Sajnekhali beat office and Pakhiryala, South 24 Parganas

Plate 3.23 A local group of folk dancers entertaining the tourists in the evening, Sajnekhali, South 24 Parganas

full-time, year-round livelihood option. Moreover, tourist spots are not evenly dispersed across Sundarban (Map 3.1). They are concentrated in few pockets, and lack coordination among them. As a result, a small proportion of people get the opportunity to be involved in tourism. Meanwhile, the tourists are not able to visit all the tourist spots in a single trip.

A proper policy to mobilize the local people is the key to revitalizing eco-tourism in Sundarban. The people residing along the fringe (1.0–1.5 km. away) of the forest have already taken part in the activities of the Eco-Development Committees (EDCs) and Forest Protection Committees (FPCs) by the Joint Forest Management

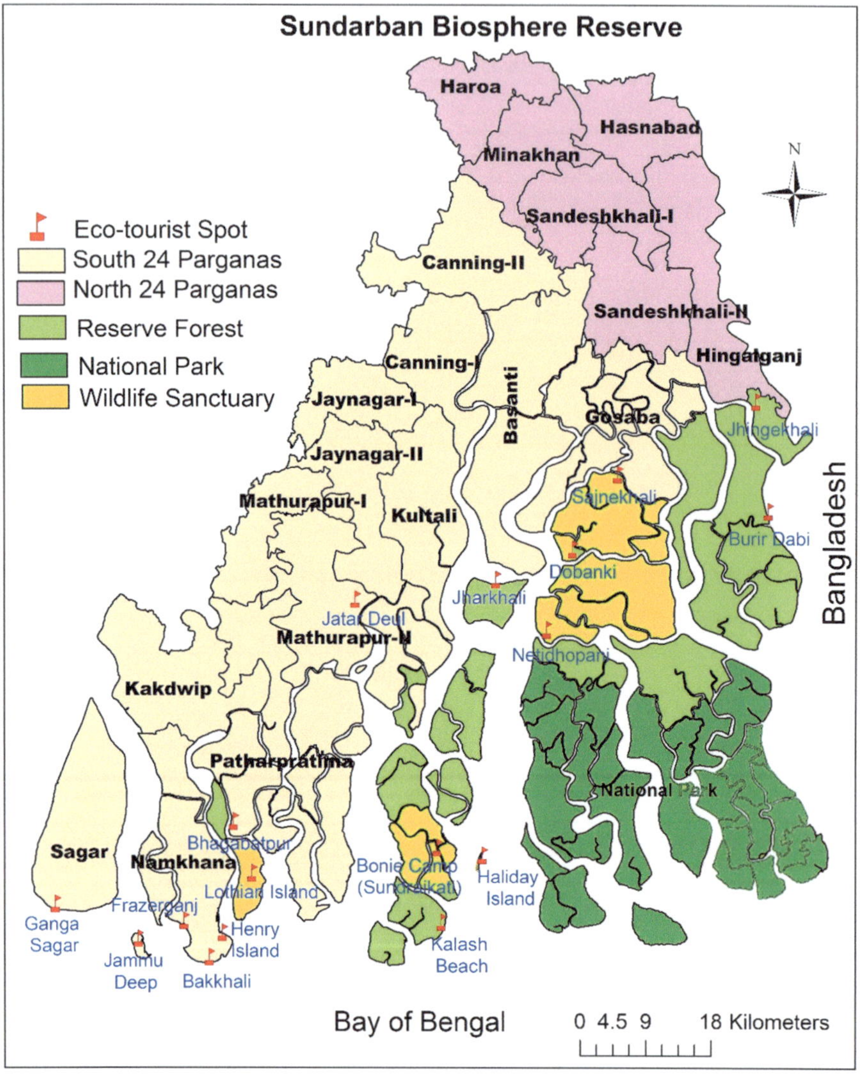

Map 3.1 Eco-tourism spots in Sundarban Biosphere Reserve, India (Source: Developed by Authors)

Eco-Tourism Spots:

Sajnekhali, Dobanki, Netidhopani, Jhingekhali and Burir Dabi in STR, Jharkhali, Bonie Camp (Sundraikati), Haliday Island, Kalash Beach, Bhagabatpur Crocodile Project, Lothian Island Sanctuary, Bakkhali Sea Beach, Frazerganj Sea Beach and Fishing Harbour, Jammu Deep, Henry Island Fishery Project, Jatar Deul in Raidighi and Ganga Sagar.

Committee of the State Government. EDC- and FPC-sponsored self-help groups are also operating along the fringe areas. Ecotourism could be integrated with them to ensure the greatest possible level of community participation. Homestay-type accommodations should be encouraged among the community. Ultimately, the adoption of a pro-people policy to retain the majority of revenue generated from tourism among the local people is very essential. Other considerations include the possibility of environmental threats, such as damage to mangrove vegetation and loss of species; riverine pollution from sewage and oil spills from mechanised boats and launches; air pollution from the emissions of boats, launches, and engine vans; and the intensification of natural calamities.

Finally, it can be concluded that the people of Sundarban face a variety of challenges that jeopardize their activities related to their principal means of living—farming and fishing. Farming suffers from seasonality, inadequate irrigation facilities, and other infrastructural issues. Climatic events and processes are also constant threats in farming. Fishing is restricted by the environmental legislations. The principal activities, therefore, have gradually been diverted as sustainability that pushes the islanders into marginality and high rates of unemployment. Ecotourism could be the best alternative at this juncture. However, the local people show extreme reluctance toward ecotourism due to a lack of incentives in terms of revenue generation and distribution among them, seasonality, and a skewed dispersion of the tourist spots. In these circumstances, migration is often the best option to cope with the situation.

References

Agricultural Census. (2000-01, 2005-06, 2010-11 & 2015-16). Survey data. Department of Agriculture, Cooperation and Farmers Welfare, Ministry of Agriculture and Farmers Welfare, Government of India (GOI). http://agcensus.dacnet.nic.in/. Accessed 28 June 2018.

B. A. E. & S. (2004, 2006, 2007, 2008, 2011 & 2013). District *statistical handbook, South 24 Parganas and North 24 Parganas*. Department of Statistics and Programme Implementation, The Government of West Bengal. Kolkata, New Secretariat Buildings: Bureau of Applied Economics and Statistics (B.A.E. & S). http://www.wbpspm.gov.in/publications/District%20Statistical%20Handbook. Accessed 22 Nov 2018.

Bhattacharya, M., Bhattacharya, P., & Patra, B. (2011). SWOT analysis of ecotourism in the Sundarban, West Bengal, India. http://philica.com/display_article.php?article_id=236. Accessed 3 July 2015.

Census of India. (1991, 2001 & 2011). *Census tables*. Office of the Registrar General and Census Commissioner, Ministry of Home Affairs, Government of India. CD ROM.

DISHA. (2006). Fishing community issues in the Sundarban Tiger Reserve (STR). The society for direct initiative for social and Health action (DISHA) and The international collective in support of fish workers (ICSF). http://aquaticcommons.org/2078/1/sundarbans_report_2march_kg.pdf. Accessed 5 July 2012.

DISHA. (2009). Traditional fishers in the Sundarban Tiger Reserve: A study on livelihood practice under protected area. The society for direct initiative for social and health action (DISHA) and the international collective in support of fish workers (ICSF). http://www.dishaearth.org/Traditional%20Fishers%20in%20The%20Sundarban%20Tiger%20Reserve.pdf. Accessed 5 July 2012.

DoSA. (2018). Sundarban Development Board. Department of Sundarban Affair, Government of West Bengal. http://www.sundarbanaffairs.in/about_us.php . Accessed 6 June 2018.

Even-Paz, Z. and Shani, J. (1989). The Dead Sea and Psoriasis: Historical and Geographic Background. International Journal of Dermatology, 28 (1), 1-9.

Ghosh, A. (2012). *Living with changing climate: Impact, vulnerability and adoption challenges in Indian Sundarbans*. New Delhi: Centre for Science and Environment (CSE). http://cseindia.org/userfiles/Living%20with%20changing%20climate%20report%20low%20res.pdf. Accessed 20 Dec 2013.

Hazra, S., Samanta, K., Mukhopadhyay, A., & Akhandet, A. (2010). *Temporal change detection (2001–2008) study of Sundarban*. Final report. School of Oceanographic Studies, Jadavpur University. http://www.iczmpwb.org/main/pdf/ebooks/WWF_FinalReportPDF.pdf. Accessed 23 Jan 2013.

Jalais, A. (2011). *Forest of tigers: People, politics and environment in the Sundarban*. New Delhi: Routledge Taylor and Francis Groups.

Kothari, A., & Patel, A. (2006). *Environment and human rights: An introductory and essential readings*. New Delhi: National Human Rights Commission.

Mistri, A. (2012). *Natural resource dependence and migration from Sundarban biosphere reserve: Case study of a remote Island*. The Centre for the Study of Regional Development, Jawaharlal Nehru University, (Unpublished Dissertation).

Mistri, A., & Das, B. (2018). Forest conservation and restraints on economic freedom of fishermen in Sundarban Tiger Reserve, India. *Indian Journal of Human Development, 11*(3), 402–409.

Sen, A. (2016). *Development as freedom*. New Delhi: Oxford University Press.

SOES. (2010). *Groundwater arsenic contamination in West Bengal – India (20 years study)*. http://www.soesju.org/arsenic/wb.htm. Accessed 26 Mar 2018.

UNO (2015). *Universal declaration of human rights (UDHR)*. United Nations. http://www.un.org/en/udhrbook/pdf/udhr_booklet_en_web.pdf, Accessed 23 June 2017.

Migration from the Sundarban

4

Abstract

Migration is a win-win solution in the Sundarban; it helps to the economic development in one hand and forest conservation by reducing the pressure on natural resources on the other hand. Short circularity around the year is the unique characteristic of migration pattern in the Sundarban, which is conducive for the regular flow of remittances. Circularity is highly associated with the seasonality of the principal activities. Sundarban people more or less equally prefer interstate and intrastate migration, but nearest urban centre, Kolkata UA, exerts a huge pull and all-time best option to the migrants. This chapter includes a detailed discussion of the processes and pattern of migration in the Sundarban.

Keywords

Migration stream · Circular migration · Seasonal migration · Remittances

Migration is a barometer of changing socioeconomic and political condition of a region. In the Sundarban, migration is a prominent livelihood strategy (Mistri and Das 2018). At least one member from three-fourth of the households migrates for seeking job (Mistri 2012, 2013). The lack of job opportunities and low wage rate are the key drivers of migration from the Sundarban. During the mid of the 2000s to mid of the 2010s, there was a trend of interstate migration from the Sundarban that surpassed the other streams, such as interdistrict and intra-district migration. The interstate migration was highly moderated by the intermediaries (Mistri 2012, 2013). In the present chapter, the processes of migration like the pattern, streams and circularity, their background characteristics and remittance flow are investigated with empirical rigour. It has been prior mentioned that data limitation is a barrier for migration study in the Sundarban as it is not an administrative area rather a functional region designed by the UNESCO. Hence, filed investigation is the prime source of data for the study.

A. Mistri, B. Das, *Environmental Change, Livelihood Issues and Migration*,
https://doi.org/10.1007/978-981-13-8735-7_4

4.1 Snapshot of Filed Investigation

Total 400 respondents are selected from 400 households, in which 200 households are defined as migrated households (from where 200 migrants are selected) and rest are non-migrated households (Table 4.1). Total 252 people equivalent to 26% of the total population move out of migrated households. That means one in every three is a migrant in the migrated households.

Average household size of migrants is higher (4.81) than the non-migrants (4.51). It has been observed that large family diversify their labour force portfolio by sending some member out and taking the risk of migration (Massey 1990; Stark and Lucas 1988). The total enumerated population from 400 households is 1,858 people. The overall sex ratio is 960 females per 1000 males. The sex ratio among the migrated households is 967 females per 1000 males and 952 females per 1000 males among non-migrated households. The mean age of migrants and non-migrants are 21 years and 25 years, respectively. About 55% are scheduled caste (SC), 15% are scheduled tribe (ST), 7.0% are other backward class (OBC), and the rest (23%) belong to general caste. Nearly 88% respondents believe in Hinduism followed by the Islam (8.0%) and Christian (3.0%). Respondents are male dominated. Most of the respondents are currently married (86%), 13% are unmarried, and the rest are widow.

Table 4.1 Snapshot of the field survey

Individual statistics			
Attributes	Migrants	Non-migrants	Total
Respondents	200	200	400
Sex	Male, 192; female, 8	Male, 178; female, 22	Male, 370; female, 30
Mean age of migration	21 years	25 years	23 years
Caste	SC, 57%; ST, 19%; OBC, 6.0%; others, 18%	SC, 54%; ST, 11%; OBC, 8.0%; others, 27%	SC, 55%; ST, 15%; OBC, 7.0%; others, 23%
Religion	Hindu, 91%; Muslim, 8%; Christian, 1%	Hindu, 86%; Muslim, 8%; Christian, 6%	Hindu, 88%; Muslim, 8.0%; Christian, 3.0%
Marital status	Married, 81%; unmarried, 20%	Married, 92%; unmarried, 6%; widow, 2%	Married, 86%; unmarried, 13%; widow, 1%
Household (HH) statistics:			
Attributes	Migrated HH	Non-migrated HH	Total HH
Enumerated HHs	200 HHs	200 HHs	400 HHs
Mean HHs size	4.81	4.51	4.66
Population and sex	Male, 487; female, 471; total, 958	Male, 461; female, 439; total, 900	Male, 948; female, 910; total, 1858
Sex ratio	967 females/1000 males	952 females/1000 males	960 females/1000 males

Source: Field survey data

4.2 Migration Streams

Previous studies (Mistri 2012, 2013) stated that people of the Sundarban preferred interstate migration (around 64%) over interdistrict (32%) and intra-district (3.0%) migration. The present study reveals that interstate migration (49%) has very slightly been preferred to interdistrict migration (48%). Only 4.0% migrants have preferred to intra-district migration (Table 4.2). The previous studies were carried out just after the severe cyclonic storm, Aila (May 2009). Aila severely affected the livelihood opportunities in the Sundarban. A huge number of distressed people were forced to move out of the Sundarban, especially interstate migration through the intermediaries. Nearly about 7 years have been passed after the severe cyclonic storm. The situation has gradually been improved. The rate of migration is still high, but the difference between interstate and interdistrict migration has reduced. Individual preferences for destination show that Kolkata ranks top followed by Tamil Nadu (13.0%), Karnataka (7.0%), Kerala (6.5%), Gujarat (6.0%) and Bardhaman (5.0%) (Fig. 4.1).

Kolkata urban agglomeration (UA) is the all-time best option to the migrants as it is adjacent to the SBR and distance varies from 35 to 125 km. Three South Eastern Railway networks, such as Sealdah-Canning, Sealdah-Diamond Harbour and Sealdah-Namkhana, and road networks connect Kolkata to the interior part of the Sundarban. Average train availability is at every 15 min. Even, some of the islanders commute 1.30–3.00 h to work in Kolkata. Another interdistrict destination, Bardhaman, is characterised by seasonal pull. Bardhaman is the 'rice bowl' of Bengal. Nearly 78% migrants, those who preferred to migrate to Bardhaman, reported that rice harvesting, especially in winter season, was the main attraction, and all of them had good contact with landowners before the move. It is a long-standing practice. Now it is facilitated by telephone and mobile network. Until the 2000s, one member from a group of 5–6 workers was sent to the work site to bring information regarding workload, number of workers required, wage rate, accommodation, etc. After that, they would make a group and set journey. Sometimes, they communicated with the landowner through latter; in response to that, they would make a plan for seasonal move.

On the other hand, a substantial proportion of migrants prefer to move out of the home state, West Bengal. Wage rate differences and job opportunities at the place of destinations are the major pull factors for interstate migration; around 49% and 37% migrants advocated the facts, respectively (Table 4.3).

Table 4.2 Migration streams from the Sundarban

Streams	Migrants	
	Frequency	Percent
Intra-district	8	4.0
Interdistrict (intrastate)	95	47.5
Interstate	97	48.5
Total	200	100.0

Source: Field survey data

Fig. 4.1 Migrants' place of destination

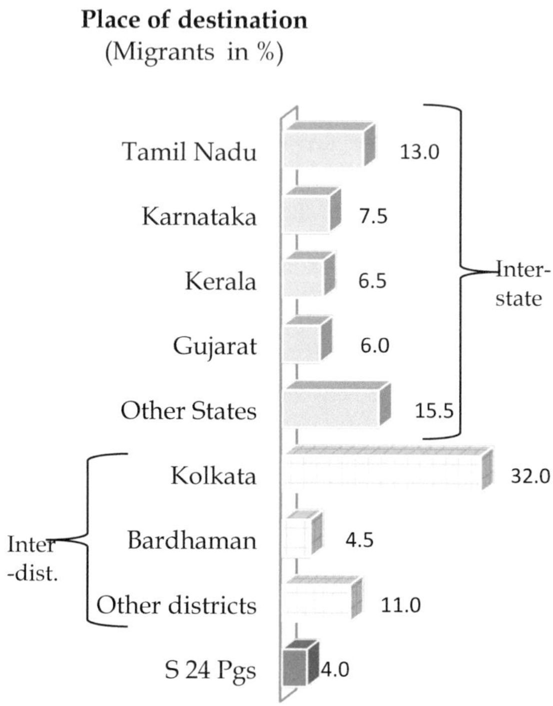

Table 4.3 Reasons for preference by streams

Preferences	Average preference 1st & 2nd (%)			
	Intra-district	Interdistrict	Interstate	Total
Greater job opportunities	31.3	36.8	37.1	36.8
Higher wage rate	56.3	44.7	48.5	47.0
Better health facilities and other basic amenities/ more comfortable life	0.0	4.3	0.5	2.3
Relatives/family members/friends shifted before	0.0	4.8	3.1	3.8
Higher education/academic facilities	0.0	2.7	1.0	1.8
Higher scope of business	6.3	4.3	0.0	2.3
Wandering new places and learning new	0.0	0.0	4.7	2.3
Others	0.0	1.6	2.6	2.0
Do not know	6.3	1.1	2.6	2.0
Total	100.0	100.0	100.0	100.0

Source: Field survey data

Furthermore, interstate migrants are well informed regarding the job availability at the place of destination (Table 4.4) that reduces the insecurity in the labour market. Around each of 30% interstate migrants received information from relatives or friends and co-villagers or neighbours. Nearly 34% of migrants are observed well-

Table 4.4 Networking used to find jobs at destinations

Sources	Intra-district migrants		Interdistrict migrants		Interstate migrants		Total	
	N	Percent	N	Percent	N	Percent	N	Percent
Relatives/friends	1	12.5	38	40.0	29	29.9	68	34.0
Co-villagers/neighbours	5	62.5	25	26.3	29	29.9	59	29.5
Middlemen/contractor	1	12.5	11	11.6	33	34.0	45	22.5
Contact himself/contact by employers	0	0.0	14	14.7	3	3.1	17	8.5
Others	1	12.5	7	7.37	3	3.09	11	5.5
Total	8	100.0	95	100	97	100	200	100

Source: Field survey data

attached with the middlemen or contractor (locally called *'thikadar'*), and they migrate to take up a job at the destination. Intermediaries contact with the workers and fix the wage rate, make a group and take them to the work site. Therefore, migrants are cent percent informed about the expected wage rate and job availability. Not a single day is wasted on the wait. Operations of middlemen reveal that they also give Rs. 1000–2000 cash in advance for the family expenditure, though it is deducted later on from the migrants' wages. In some cases, the transportation cost is borne by the contractor, especially one side travel cost to go to the work site. Over the time, some migrants also turn into contractors.

4.3 Migration Pattern

The circular pattern is characterised by the migration from the Sundarban. Circular migration is defined as the repeated movement and return between an origin and destination, and the repetition must be more than one (Hugo 2013). In the Sundarban, nearly 98% migrants have experienced return migration, of which around 59% have experienced more than one-time return during the last 365 days, and the rest (41%) have experienced only one-time or yearly return (Table 4.5). The mean number of returns is seven times during the last 365 days. Moreover, those who returned, around 84% of them, had preferred the same place of destination to their last move. But, the rest, 16%, changed their destinations, and they are comparatively young – around 73% are below the 35 years. Hence, they took the risk to migrate a new place of destination in every return. According to Graeme Hugo's (2013) definition, in the Sundarban, around 62% of migrants have experienced more than one return to origin and movement to the same place of destination. Within 62%, 31% migrants have returned for 2–4 times, 5% have returned for 5–11 times, and 26% have experienced monthly, fortnight and weekly return. These 26% is better to be defined as mobility. Frequent return and migration are associated with short-distance movement, such as interdistrict, intra-district and seasonal movement, where migrants have well acquaintance with the destination and its job markets.

Table 4.5 Migrants by number of return and place of last move

No. of return during the last 365 days	If return, migrated to the same place of destination in the last move?					
	Yes		No		Total	
	Frequency	Percent	Frequency	Percent	Frequency	Percent
1 time/yearly return	79	38.3	21	52.5	100	40.7
2–4 times return	63	30.6	14	35.0	77	31.3
5–11 times return	11	5.3	2	5.0	13	5.3
Monthly, fortnight and weekly return	53	25.7	3	7.5	56	22.8
Total	206	100.0	40	100.0	246	100.0

Source: Field survey data

On the other hand, one-time return pertains to the long-distance migration, especially interstate migration. Due to huge migration cost, such as monetary, opportunity and psychic costs, interstate migrants often avoid the frequent return. Moreover, interstate migration is highly moderated by the intermediaries. The migrants often change their contractors according to their suitability. Hence, with the changes of the contractors, work sites are changed or vice versa.

4.4 Circularity and Its Characteristics

Agunias and Newland (2007) classified the circular migration into four categories on the basis of the nature of departure and return of migrants. Both the departure and return may be either permanent or temporary. They cited four distinctive country-specific studies from different time periods and divided the circular migration into four categories (Table 4.6).

According to their typology, the migrants of the Sundarban are a temporary return of temporary migrants. The workers neither permanently migrate from the Sundarban nor permanently return. They return home only to migrate again. In the Sundarban, around 28% migrants reported that they will permanently return within 2 years (Table 4.7). Whereas, more than two-third migrants (72%) opined that they do not know when they will permanently return, but they are very sure that they will not stay at the destination permanently. In this context a verbatim of a construction labour named Mr. Provash Mondal mentioned:

> *Who would like to stay away, leaving home and family? Migrate for the stomachs. In a few days, earn a good amount. Now age permits for the work; when it does not, then will not go.*

However, a close tie-up with the origin (or family) is observed among the migrants resulting in circulation of migrants, and circularity is conducive to the remittance flow as well as economic development of an area.

It has already been discussed in the previous chapter that principal activities, farming and fishing, in the Sundarban suffer seasonality. The seasonality in occupation is one of the leading factors for circularity of migrants around the year (Fig. 4.2).

Table 4.6 Typology of circular migration

	Permanent migrants	Temporary migrants
Permanent return	**A.** Return of the Irish diaspora in the late 1990s	**B.** Korean turnkey project manager in the Middle East
Temporary return	**C.** Taiwanese "astronauts" from Canada and Silicon Valley, California	**D.** Contract workers from the Philippines

Source: Based on Agunias and Newland (2007, p. 4)

Table 4.7 Intention to return permanently in the Sundarban

Duration	Frequency	Percent
Less than 6 months	24	9.5
6 months–2 years	47	18.7
Do not know	181	71.9
Total	252	100.0

Source: Filed survey data

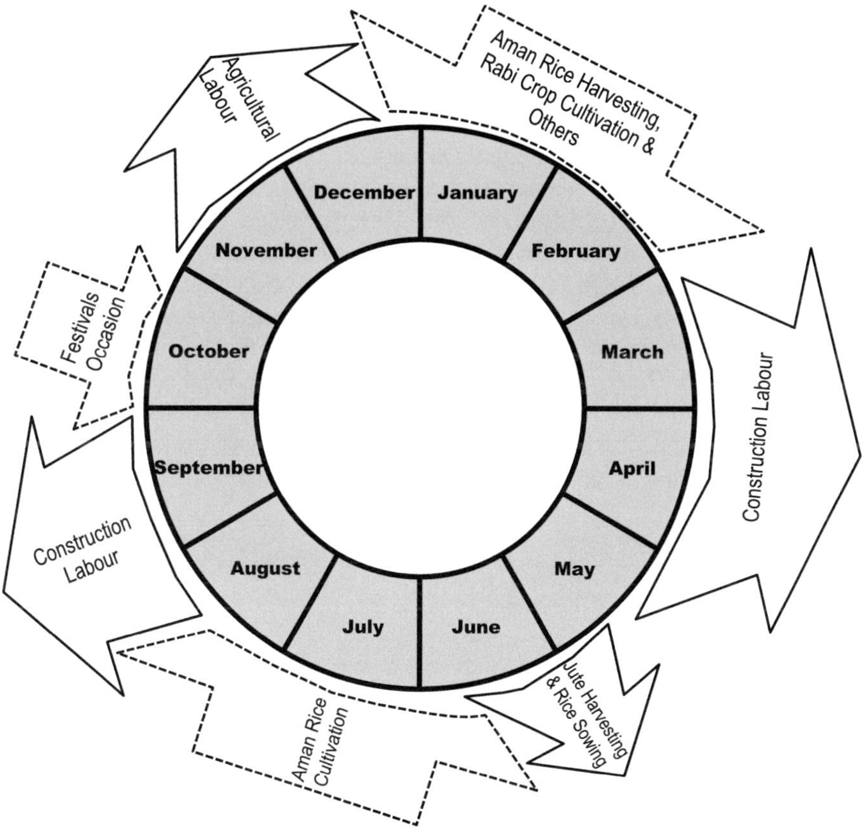

Fig. 4.2 Annual pattern of migration in the Sundarban

From mid-June to mid-August is the full agricultural season in the monsoon period in the Sundarban. Preparation for Aman rice cultivation is taken place in a full swing. After that, almost no work is available at a large scale until the Aman rice is ripe and harvested. During this lean period, people migrate out of the Sundarban, either in Kolkata and other districts of West Bengal or out of the state for construction work. September and October are the festival months of Bengali community. Durga Puja, Kojagari-Laxmi Pooja, Diwali or Kali Puja, Bhai Dooj, Eid al-Adha, Maharam, etc. are major auspicious occasion during the months. Migrants return for a short span to celebrate his/her respective festivals and spend a valuable time with family members after a long alienation. From mid-November to mid-December (around 1–-1½ months), a seasonal interdistrict migration stream is observed. The workers move in a group for agricultural labour, such as paddy harvesting and sowing of potato and other Rabi crops, in district of Hooghly, Howrah, Bardhhaman, Birbhum and East Medinipur. These districts are the well irrigated, and fertile lands are practised for intensive agriculture.

In between mid-December and mid-February, migrants return back to harvest Aman rice at the place of origin. The agricultural land in the Sundarban is low-lying. Hence, drying up and preparing the land for harvesting of Aman rice take time and follow a late harvesting season compared to the other districts of West Bengal. As a result, workers get the opportunity to be engaged as agricultural labourer in both places. From March to June, a small section of farmers who have enough sources of irrigation practice Rabi crop. Besides, it is a dry season, and some job opportunities are also created at daily labour basis such as road repair, pond and canal digging and renovation, embankment repairing, etc. But, it is not emerged at a large scale, limited to some pockets and demand basis. Moreover, 3 months, namely, April, May and June, are the closed season for fishing. Hence, they again obligated to move out. During June to July, a very small section of people migrate in a group for jute harvesting and replace the same field with Aman rice in Hooghly, Howrah, Nadia and Murshidabad. Thus, workers in the Sundarban move out and come back more than three times in a year, and it is related to the seasonality in principal activities. Sometimes, the short circularity of migration in the Sundarban is broken down by the occurrence of natural hazards and takes time to restore roughly 4–5 years.

4.5 Reasons for Migration

The reasons for migration (Table 4.8) reveal that most of the migrants, that is, 88%, have moved out of the Sundarban for seeking a job, of which 43% have migrated 'in search of employment', 33% for 'to take up employment or better employment' and 13% for 'in search of better employment' (Appendix 2). Nowadays, the modern form of transportation and communication network significantly boosts up the circular migration. The Sundarban is not its exception. Every nook and corner of the islands have access to telephone connections, and growing rail and road network are trying to attach the interior part to the mainland.

Table 4.8 Reasons for migration

Reasons	Frequency	Percent
In search of employment	107	42.5
In search of better employment	32	12.7
To take up employment/better employment	83	32.9
Proximity to place of work	13	5.2
Studies	11	4.4
Others	6	2.4
Total	252	100.0

Source: Field survey data

4.6 Occupation and Migration

From the field investigation, it is observed that nearly half of the respondents have reported as marginal workers (Table 4.9), whereas main workers are around 39.3% and nonworkers are 11.3%. Most of the migrants (54%) were marginal workers engaged less than 3 months in the work at the origin during the last 365 days before migration, whereas around 45% non-migrants are marginal workers. The main workers (51%) are comparatively higher among the non-migrants, but a substantial proportion of marginal workers (45%) are also observed among them. Moreover, nearly 18% of the migrants have reported that they were not working at the origin prior to their move. A chi-square ($x^2 = 30.73$, $df = 2$, $n = 400$) test for independence (with Pearson chi-square) indicates that there is a significant association ($p = 0.00$) between the work status and migration status of respondents (migrants or non-migrants). The degree of association between the variables is moderate (Cramer's V = 0.30). The adjusted residuals (Adj.R) reveal that both the main worker (Adj.R = −/+ 4.6) and nonworker (Adj.R = +/−4.3) are significant contributors to the x^2 test model (Feature Box 4.1). The main worker status of labourer strongly advocates retaining at the place of origin. On the other hand, the nonworker status highly argues in favour of migration from the Sundarban.

Feature Box 4.1: Standardised Residual and Adjusted Residual in x^2 Test

The residual in chi-square (x^2) test is computed from the difference between observed cell count and expected count (Observed Cell Count – Expected Count). Greater the residual produces greater x^2 value, and more likely to be rejected the null hypothesis (H_0) that is two variables are independent or not associated. The p value ≤ 0.05 rejects the null hypothesis with 95% level of confidence.

(continued)

Feature Box 4.1 (continued)

Adjusted Residual (Adj.R) = (Standardised Residual ÷ Estimated Standard Error);

Where, Standardised Residual = (Observed Cell Count − Expected Count) ÷ √ Expected Count.

For the 5% level of significance, the critical value for an adjusted residual is −1.96 and + 1.96 or round off +/− 2. A cell (in a crosstab) with adjusted residual greater than the cut-off point 2.0 would be deemed to be significant contributor to make difference between two variables.

Table 4.9 Work status of respondents during the last 365 days

Types of workers	Attributes	Migrants' work status before moving out during the last 365 days	Non-migrants' work status during the last 365 days	Total
Main worker	Count	56	101	157
	Expected count	78.5	78.5	157
	% within workers	**28.0%**	**50.5%**	**39.3%**
	Adjusted residual	−4.6	4.6	
Marginal worker	Count	108	90	198
	Expected count	99	99	198
	% within workers	**54.0%**	**45.0%**	**49.5%**
	Adjusted residual	1.8	−1.8	
Nonworker	Count	36	9	45
	Expected count	22.5	22.5	45
	% within workers	**18.0%**	**4.5%**	**11.3%**
	Adjusted residual	4.3	−4.3	
Total	Count	200	200	400
	Expected count	200	200	400
	% within workers	**100.0%**	**100.0%**	**100.0%**

Source: Computed from field survey data

4.7 Work Engagement of Migrants: Before and After Move

The job inadequacy, which leads to marginality (work less than 6 months), is sharply visible in the Sundarban (Table 4.9). Moreover, when this marginality highly persists in the principle means of livelihoods, the crisis becomes worse that leads to migration by leaps and bounds. More than one-third of the migrants (38%) have reported that they were engaged in farming before the move (Table 4.10). Likewise, fishing and daily or casual labourer accounts for 15% each. A fair proportion of migrants (9.0%) are forced to leave their study in middle and secondary standard to enter the job market. Another 10% nonworkers have experienced migration for the first time.

More than half (52%) of the migrated person are working as a construction labourer at the place of destinations. Around 10% migrants have reported as agricultural labourer engaged in seasonal harvesting. The rest are porter in storage and dockyard, helper in the truck (*khalasi*), rickshaw puller, hawker in train or bus, nurse (*Aya*) and maidservant. Some migrants are engaged as industrial labourers in the garment factory, such as tailor, rice mill, iron-grill industry, goldsmith, leather industry and manufacturing industry.

Table 4.10 Migrants' work engagement before and after moving (in absolute figure)

	Work engagement at destination										
Work engagement before moving		**i**	**ii**	**iii**	**iv**	**v**	**vi**	**vii**	**viii**	**ix**	**Total**
	i	5	0	7	47	6	4	3	4	0	**76**
	ii	4	0	5	19	0	1	1	0	0	**30**
	iii	9	1	5	12	2	0	0	1	0	**30**
	iv	0	0	0	3	1	0	0	0	0	**4**
	vi	0	0	0	2	0	1	0	0	0	**3**
	vii	0	0	0	1	0	0	2	0	0	**3**
	viii	1	1	0	1	0	0	0	2	0	**5**
	ix	0	0	1	7	1	2	1	1	5	**18**
	x	0	0	4	2	3	1	0	1	0	**11**
	xi	1	0	3	9	5	2	0	0	0	**20**
	Total	**20**	**2**	**25**	**103**	**18**	**11**	**7**	**9**	**5**	**200**

Note: **i**, Farming; **ii**- fishing and NTFPs; **iii**, daily labour/casual labour; **iv**, construction labour; **v**, manufacturing/industrial labour; **vi**, regular wage/salary earning; **vii**, business; **viii**, transport and communication; **ix**, study; **x**, household industry; **xi**, nonworker

Source: Field survey data

4.8 Remittances

Importance of remittance is inevitable in the development process, where the resources are constraint in the economy and saving is suboptimal. Almost US$ 318 billion remittances are remitted by the migrants worldwide, and 80% of which flow towards developing countries (Hoermann and Kollmair 2009). Apart from financial accounting, it has immense social value. Social remittances open the window to the modernisation of the rural or backward regions. Remittances are two broad types – financial and social remittances (Banerjee et al. 2011). Financial remittances are further divided into monetary and nonmonetary remittances. Social remittances include the skills, ideas, practices and knowledge that migrants carry back to their origin communities. The aspects related to financial remittances, such as amount and frequency of remittances flow, and its utilisation are discussed within the limited scope of the study.

More than half of the migrants (59%) sent money to their family during the last 365 days. Around 56% migrants send money for 1–3 times, 20% transferred for 4–11 times and 24% remitted frequently that is 12 or more times in a year. The average frequency of remitting is five times.

The amount received by the households in the Sundarban widely varies from a minimum of Rs. 1000 to maximum of Rs. 100,000 per year. The standard deviation is too high (Table 4.11). The median of remittance is Rs. 8500 per year. A substantial proportion of migrants (42%) have reported that they have remitted Rs. 15,000 and above during the last 365 days. Nearly 19% migrants have sent in between Rs. 5,000 and Rs 15,000 and the rest, 39%, have remitted Rs. 5000 and below during that period. Therefore, a fair amount of remittances regularly flows into the Sundarban.

Spending on food items is the uttermost importance in the expenditure list of the migrated households. On an average, 42% households spend for the same (Table 4.12). It is followed by spending for the education (16%) of household members and improving housing condition (10%). The effort of saving and investment in the entrepreneur and making financial working capital are very minimal. A sheer amount expense for bread and butter and education of children are an effort of leaving out of poverty and backwardness. The frequent flow (on an average five times) of remittances with a fair amount make the pave widens.

There is a wide debate among the many policymakers in the recent time that circular migration is a 'triple-win' solution or a 'zero-sum game' (Angenendt et al.

Table 4.11 Amount (Rs.) of remit in the Sundarban

N	117	Amount (Rs.)	Migrants	Percent
Mean	Rs. 17,389	Rs. 5,000 and below	46	39.3
Median	Rs. 8,500	Rs. 5000–15,000	22	18.8
Mode	Rs. 5,000	Rs. 15,000 and above	49	41.9
Std. deviation	19,363	Total	117	100
Range	Rs. 99,000			
Minimum	Rs. 1,000			
Maximum	Rs. 100,000			

Source: Field survey data

Table 4.12 Usages of remittances, average preference (1st and 2nd)

Usages	HHs	Percent
Food items	49	41.5
Education of household members	19	16.2
Improving housing condition	12	10.3
Household durable	7	5.6
Financing working capital	7	5.6
Others	25	20.9
(Debt repayment, other items on household consumer expenditure, healthcare, initiating new entrepreneurial activity, saving/investment)		
Total	117	100

Source: Field survey data

2015; Angenendt 2014; Wickramasekara 2011). Triple-win solution of circular migration implies bringing benefits to destination countries, origin countries and migrant workers themselves. Studies (Hugo 2013; Agunias and Newland 2007) suggest that the circular migration is one of the few means for resources to flow from the core areas like cities and high-income areas or countries to the peripheral areas, such as rural areas and low-income areas or countries. As a result, it reduces poverty and inequality and fosters the developments in the backward regions. These benefits are maximised when migration occurs in a 'favourable policy context' and migrants retain strong bonds to their place of origin and make frequent return trips (Hugo 2013). In the Sundarban, when agriculture and fishing are restricted by different physical and environmental factors, circular migration is the key to the economic development. It helps to the economic development in one hand and forest conservation by reducing the pressure on natural resources on the other hand. The islanders have strong ties to their origin and make frequent return trips and remigrate. A fair amount of remittances is regularly remitted to the family. But there is a lack of coherent policy framework for the social protection of the migrants; it is not only observed in the Sundarban but also in India (Srivastava 2012). As a result, migrants suffer from lots of misery, work hazards and trafficking at the destination. Moreover, there is a lack of enough investment opportunities of their remittances in the Sundarban.

References

Agunias, D. R., & Newland, K. (2007). *Circular migration and development: Trend, policy routes, and ways forward*. Washington DC: Migration Policy Institution (MPI).

Angenendt, S. (2014). *Triple-win migration – Challenges and opportunities*. The German Marshall Fund of the United States (GMF) and the Robert Bosch Stiftung. http://www.gmfus.org/publications/triple-win-migration-challenges-and-opportunities. Accessed 28 June 2015.

Angenendt, S., Bither, J., & Ziebarth, A. (2015). *Creating a triple-win through labour migration policy?* Lessons from Germany; A report of the migration strategy group 2013–2014. The German Marshall Fund of the United States (GMF) and the Robert Bosch Stiftung. http://www.gmfus.org/publications/creating-triple-win-through-labor-migration-policy-lessons-germany. Accessed 28 June 2015.

Banerjee, S., Gerlitz, J. Y., & Hoermann, B. (2011). *Labour Migration as a Response Strategy to Water Hazards in the Hindu Kush-Himalayas*. Kathmandu: ICIMOD.

Hugo, G. (2013). *What we know about circular migration and enhanced mobility*. Washington DC: Migration Policy Institution (MPI).

Hoermann, B. & Kollmair, M. (2009). Labour migration in the Hindu Kush-Himalayas – A core livelihood strategy. ICIMOD, Kathmandu. http://lib.icimod.org/record/26470/files/attachment_594.pdf

Massey, D. S. (1990). Social structure, household strategies, and the cumulative causation of migration. *Population Index, 56*(1), 3–26.

Mistri, A. (2012). *Natural resource dependence and migration from Sundarban Biosphere reserve: Case study of a remote Island*. The Centre for the Study of Regional Development, Jawaharlal Nehru University, (Unpublished Dissertation).

Mistri, A. (2013). Migration and sustainable livelihoods: A study from Sundarban biosphere reserve. *Asia Pacific Journal of Social Sciences, 5*(2), 76–102.

Mistri, A., & Das, B. (2018). Migration in response to environmental change: A risk perception study from Sundarban biosphere reserve. In S. Irudaya Rajan & R. B. Bhagat (Eds.), *Climate change, vulnerability and migration*. New Delhi: Routledge.

Srivastava, R. (2012). Internal migration in India initiative. In *National workshop on internal migration and human development in India*, New Delhi, India, 6–7 December 2011. New Delhi: United Nations Educational, Scientific and Cultural Organization (UNESCO).

Stark, O., & Lucas, R. E. B. (1988). Migration, remittances, and the family. *Economic Development and Cultural Change, 36*(3), 465–481.

Wickramasekara, P. (2011). *Circular migration: A triple win or a dead end*. Geneva: International Labour Office.

Environmental Risk Perception and Migration

5

Abstract

Environmental migration occurs when the environment adversely affects a person's livelihood, resulting in migration as an outcome. Simply put, migration is linked to environmental stressors through livelihood issues. This chapter examines risk perception study, which is one of the best approaches to assess the association between livelihood issues and environmental stressors. The subjective appraisal of the environmental threats to their livelihood is the focal theme of risk perception study. The people of Sundarban have a moderately strong risk of environmental adversity in their activities, especially farming and fishing, whether they are migrants or non-migrants.

Keywords

Risk perception · Affective imagery · Psychometric paradigm · Psychophysical scaling

The association between environmental change and migration is hardly linear; rather, multiple factors need to be examined along a spectrum of socioeconomic and demographic aspects when the environment is a cause of migration (ADB 2009). Environmental migration has been defined (IOM 2009; Myers 2005; Kniveton et al. 2008) as environment-induced migration that occurs when the environment adversely affects the person's livelihood, resulting in migration as an outcome. Migration is linked to the environment through livelihood issues.

In a risk perception study, people are asked to characterize and evaluate hazardous activities (Slovic and Peters 2006; Slovic 1987). The study places emphasis on people's cognition or "affective imagery," which guides their judgments or decision making regarding the influences of stimuli on their livelihood choices (Leiserowitz 2005, 2006). Therefore, people's risk perception of livelihood can demonstrate the association between environmental change and livelihood mitigation through migration from Sundarban.

A. Mistri, B. Das, *Environmental Change, Livelihood Issues and Migration*,
https://doi.org/10.1007/978-981-13-8735-7_5

The first risk perception study was published in 1928 by L. L. Thurstone (Sjoberg 2000). The approach was more broadly accepted in the 1970s and became widely appreciated starting in the 1990s. In the twenty-first century, risk perception studies have been widely been discussed in the context of climate change in the Western world, especially in the United States (Leiserowitz 2005, 2006; O'conner et al. 1999). Studies by Leiserowitz (2005, 2006) used a mail-back survey of 673 American people to reveal that the American public perceive a moderate extent of risk from climate change. More than two-thirds (68%) of respondents were highly concerned about the impacts of climate change on people around the world and non-human nature, whereas only 13% were most concerned about the impacts on themselves, their families, or their local communities. The respondents strongly supported different sorts of national and international policies to mitigate climate change, but strongly opposed several carbon tax proposals. Nowadays, risk perception studies have been widely adopted to examine the risk perception of activities for the means of living, such as farming (Farauta et al. 2011), fishing (Edvardsson et al. 2011; Chand et al. 2012), and others. Moreover, the United Nations Framework Convention on Climate Change's Rio Convention in 1992, 1997's Kyoto Protocol, and 2009's Copenhagen Convention boosted the profile of such studies.

This chapter mainly deals with the subjective appraisal of environmental change in Sundarban, its influence on means of livelihood, especially on farming and fishing, and finally its association with out-migration. People's concerns about environmental changes in Sundarban are discussed in the first section of this chapter. The perceived risks to their activities of farming and fishing are discussed in the next section. A common approach for studying risk perception is the psychometric paradigm, which includes "psychophysical scaling," the risk perception index, areas of concern, multivariate analysis, and so on. For this study, the psychophysical scale was constructed using a Likert-type scale and Bipolar Affect Scale or Positive and Negative Affect Schedule (PANAS). In addition, risk perception indices, areas of concern, and affective imagery are also discussed. Simple statistical techniques such as mean score, standard error of mean (*SEM*), *t*-test, x^2-test, and effect size are used as computational tools.

5.1 Environmental Concern and Migration

5.1.1 Risk Perception Index

Studies on environmental parameters, which have been discussed in detail in Chap. 2, suggest that the environment in Sundarban has gradually deteriorated. Now, the questions are whether the inhabitants are concerned about the changes to the environment and whether such changes are significantly associated with the process of migration. The environmental concerns of the Sundarban people can be attributed to four categories: holistic concerns, the concern for local impacts, the concern for non-human nature, and the seriousness of current impacts around the world. These attributes help to clarify how people's concerns vary from geographically distant people and places, which changes have effects outside of Sundarban or worldwide impact, and which affect non-human nature such as mangrove forests and wild animals.

The cognitive judgements of 400 respondents were collected using different sets of questions structured with a five-point Likert-type scale (Table 5.1). Each respondent was requested to select any one of the options—namely “to a very great extent,” “to a great extent,” “to some extent,” “to a little extent,” and “to no extent.” The values assigned to these options are 5, 4, 3, 2, and 1 respectively. The mean score is 3.0 [(5 + 4 + 3 + 2 + 1)/5 classes]. The score of 3.0 is deemed to be a normal level of concern. A variable with a (sample) mean score ($\bar{x}$) less than 3.0 is considered to be a low-risk perception or low impact of that particular stimulus. A mean score above 3.0 is considered to be a great risk perception. A one-sample *t*-test was used to determine whether the sample comes from a normal population, which has a mean (μ) of 3.0. Thus, the test examines the significant difference ($p < 0.05$) between the sample mean ($\bar{x}$) and hypothesized population mean ($\mu = 3.0$). In addition, the mean difference ($\bar{x} - \mu$) expresses how much lower or higher the $\bar{x}$ is compared to μ (3.0) at a 95% confidence interval (*CI*). The standard error of mean (*SEM*) infers how precisely the sample predicts the true mean of the population or is closer to the true population mean.

The risk perception index (RPI) score of the respondents is 2.68 (Table 5.1). A one sample *t*-test (−7.645) reveals that the RPI score is significantly different ($p = 0.000$) from the hypothesized population mean ($\mu = 3.0$) at 95% *CI* (−0.40 to −0.23) and is lower by 0.315. The respondents, therefore, perceive environmental change in Sundarban as a lower risk. The mean scores of seven of the eight indicators under investigation are significantly lower than the hypothesized mean ($\mu = 3.0$). Only one indicator—that is, the chance of contracting a serious disease—is not statistically significantly different; it is equal to the $\mu = 3.0$ or average level of concern. Although the mean scores of the seven indicators are significantly lower than the hypothesized mean, they do not widely vary from the same (Fig. 5.1). The mean differences ($\bar{x} - \mu$) are not lower than 0.50, which implies that people’s environmental concern is not alarmingly low. The inhabitants of Sundarban are most concerned about the effects on their occupation ($\bar{x} = 2.85$), followed by water shortages ($\bar{x} = 2.84$) and a declining standard of living ($\bar{x} = 2.67$). They are relatively less concerned about the threats faced by non-human nature, such as mangrove forests ($\bar{x} = 2.35$) and wild animals ($\bar{x} = 2.50$), including tigers, monkeys, deer, crocodiles, turtles and tortoises, and birds. They are little concerned about the worldwide effects ($\bar{x} = 2.55$) of environmental change in Sundarban.

The study’s conclusions are reinforced by the answers to a separate question, which asked the respondents to specify which area of concern caused them the greatest anxiety due to environmental change in Sundarban (Table 5.2). A clear majority of respondents (67%) are concerned with the impacts on themselves, their families, and their local communities. A total of 25% are most concerned about the effects on Sundarban as a whole and people around the world. Only 2% are concerned about non-human nature.

Table 5.1 Environmental change risk perception index (RPI) in Sundarban

Attribute	Indicators	Mean ($\bar{x}$)	Standard error of mean *(SEM)*	Test Value: $\mu = 3$		
				Mean Difference. ($\bar{x}$-μ)	t	*p (two-tailed)*
Holistic concerns	*How concerned are you about the environmental change in Sundarban?*	2.64	0.062	−0.357	−5.793	0.000
Local impacts	*Do you think that the following incidences will occur due to the environmental change during the next 25 years?*					
	Do you think your means of living (farming/fishing/others) will be hindered?	2.85	0.065	−0.152	−2.347	0.019
	Do you think water shortages will occur where you live?	2.84	0.068	−0.160	−2.360	0.019
	Do you think your standard of living will decline?	2.67	0.052	−0.335	−6.401	0.000
	Do you think your chance of getting a serious disease will increase?	3.09	0.060	0.087	1.469	0.143
Non-human nature	Do you think the environmental changes will					
	destroy mangrove forests?	2.35	0.060	−0.648	−10.861	0.000
	Do you think the environmental changes will affect the wild animals (tigers, monkeys, deer, crocodiles, turtles and tortoises, birds, and others)?	2.50	0.059	−0.505	−8.525	0.000
Worldwide impact	How concerned are you about the current worldwide impact of environmental change in Sundarban?	2.55	0.066	−0.453	−6.876	0.000
	RPI	**2.68**	**0.041**	**−0.315**	**−7.645**	**0.000**

Scale ranges from 1 (to no extent) to 5 (to a very great extent); $n = 400$, $df = 399$, $p < 0.05$- statistically significant
Source: Computed from field survey data; based on Leiserowitz (2006)

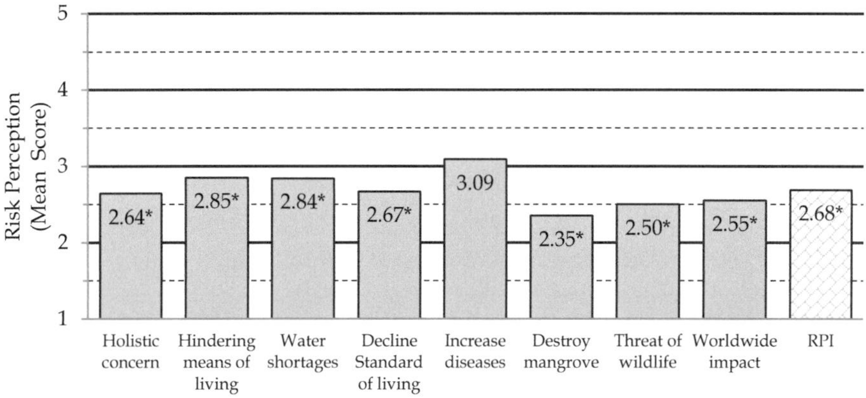

Fig. 5.1 Risk perception in the current impact of environmental change in Sundarban
Note: ∗Statistically significant ($p < 0.05$) difference from $\mu = 3.0$
(Source: Computed from field survey data)

Table 5.2 Areas of concern by inhabitants of Sundarban

Most concerned about the impact on?	Frequency	Percent	Cumulative percentage
You and your family	251	62.8	62.8
Local community	15	3.8	66.5
Sundarban biosphere reserve as a whole	60	15.0	81.5
People all over the world	40	10.0	91.5
Non-human nature	9	2.3	93.8
Not at all concerned	25	6.3	100.0
Total (*n*)	400	100	

Source: Field survey data

To understand major national or international issues, such as climate change, one should consider the direct local relevancies. Without the local people's recognition, it is impossible to address environmental problems. Nowadays, in the politics of global warming, most nations are more concerned about the carbon emissions of other countries rather than their own levels. Climate change is unlikely to become a high-priority national issue until those countries consider themselves to be personally at risk (Leiserowitz 2006).

The RPI scores for both migrants ($\bar{x} = 2.70$, $n = 200$, $SEM = 0.062$) and non-migrants ($\bar{x} = 2.67$, $n = 200$, $SEM = 0.054$) are significantly ($p < 0.05$) lower than the normal level of concern ($\mu = 3.0$); the score for migrants is slightly greater than that for non-migrants (Fig. 5.2). There is no statistically significant difference ($t = 0.416$, $df = 398$, $p = 0.677$) in the mean scores of the two groups and the magnitude of differences is very low (*Eta Squared* = 0.0004 or 0.04%). However, the migrants are more concerned about their occupation at a local level; the mean scores for migrants and non-migrants are 2.92 and 2.78, respectively, for this issue, which is one of the reasons for a decision to migrate. A declining standard

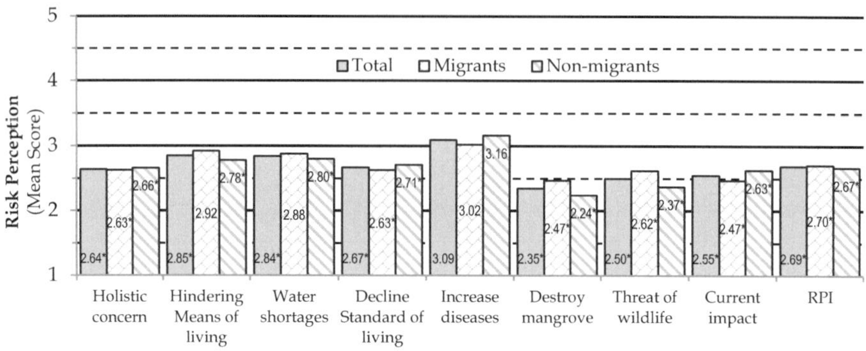

Fig. 5.2 Risk perception differences between migrants and non-migrants in Sundarban
Note: ∗Statistically significant ($p < 0.05$) difference from $\mu = 3.0$
(Source: Computed from field survey data)

of living and increased incidence of disease are slightly more concerning to non-migrants because they are firmly attached to their place of origin. Moreover, the migrants show more awareness about the environmental impact on wildlife and forests compared to the non-migrants.

5.1.2 Affective Imagery

An affective image is the combination of an image or cognitive content with an affective evaluation (Leiserowitz 2006). The "affect" is the positive or negative evaluation of a stimulus in a person's stable state (Slovic and Peters 2006). The image includes all forms of mental representations, such as perceptual representation (picture, sound, and smell), symbolic representation (word, number, and symbol), and others (Leiserowitz 2005). Affective imagery, therefore, is a mental image of a stimulus to which people's cognitive judgement is attached. To examine the affective images of climate change, two questions were asked of the respondents: the first one is related to the mental image of climate change, whereas the other is the "affect" of that image. All questions were administered in the local language, Bengali.

Q_1: What is the first thought, image, or idea that you visualize in your mind when you think of climate change in Sundarban?

If the answer to the question was a cyclonic storm, depression, or any other phenomenon, or some narrative sentences that indicate the same, then the respondent was asked the second question.

Q_2: How strong is your negative or positive feeling about [cyclonic storm, depression, or other phenomenon]?

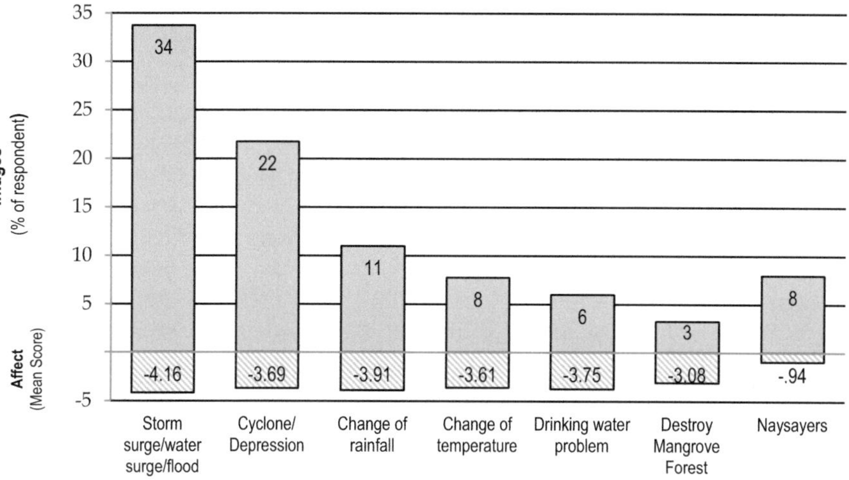

Fig. 5.3 Islander's affective images of climate change in Sundarban (Source: Computed from field survey data)

The answer was evaluated on a 10-point PANAS or Bipolar Affect Scale. It ranges from −5 to +5, namely to a very great extent (± 5), to a great extent (± 4), to some extent (± 3), to a little extent (± 2), and to no extent (± 1). The feeling of an image may be either negative (−1 to −5) or positive (1–5).

A total 14 distinct thematic categories of images was observed. Out of these 14 categories, seven predominant images were selected by 92% of the respondents (Fig. 5.3). Storm surge, water surge, or flooding is the most salient image of the impact of climate change in the minds of Sundarban people, as it was selected by 34% of the respondents. The next most important images are cyclone or depression (22%), change of rainfall (11%), and change of temperature (8.0%). Here, the cyclone or depression refers to only the wind action or storm. The wind action associated with a flood or a water surge at the time of a cyclone or depression is defined as storm surge. In Sundarban, not all depressions cause flooding, depending on the intensity, direction, and time (tidal period) of the occurrence of a cyclone. A few respondents (8.0%) reported that climate change is overhyped by the media, politicians, and so-called educated people who occasionally come to Sundarban; rather, the only problem in Sundarban is development deficit, which needs to be addressed.

The mean affect scores of all images are negative, which indicates a negative feeling regarding the impact of climate change for almost all respondents (Fig. 5.3). Although there were a few respondents who did not have clear ideas (images) regarding the consequences of climate change, they reported positive effects of the same. The image of storm surge exerts the strongest negative effect ($\bar{x} = -4.16$). Although the share of the change in rainfall (11%) is lower than the image of the cyclonic storm (22%), the former affects more (negatively) than the later one.

Likewise, the image of a drinking water problem exerts more negative effects than that of a change of temperature instead of having a proportional difference in the mental images.

The images of water surge together with a cyclonic storm imprints on more than half (56%) of the respondents' minds, and its "affect" is deeply rooted. Generally every year, Sundarban experiences tropical cyclonic storms that originate in the Northern Bay of Bengal. During the period of 2005 to 2015, a total of 12 tropical cyclonic storms formed in the Northern Bay of Bengal, including one severe cyclonic storm (SCS), Aila (May 2009) directly hit Sundarban; three other storms—Sird (super cyclonic storm) in 2007, Bijli (SCS) in 2009, and Phailin (very SCS) in 2013—immensely affected the lives of the Sundarban people (Table 2.4). Apart from the loss of life and property, these storms ruined the basic infrastructure, and people had to suffer ill health due to a lack of safe drinking water, sanitation, and medical facilities. A single flood or storm surge in Sundarban causes at least 2–3 years of complete crop failure, and an additional 2 years of low yield due to a very high level of salinity in topsoil and stored water in ponds and canals for irrigation. In addition to that, cultivation in Sundarban highly depends on the monsoon rainfall. The vagaries of monsoon rainfall thus severely influence the livelihoods of the Sundarban people. In addition, heavy rainfall often causes floods in low-lying areas of the islands.

5.2 Environmental Threats to Means of Living

Farming and fishing are affected by climatic change in different ways. Climate change may deteriorate the onset and growing environment, decrease production or collection, degrade the quality of production, increase diseases and other affecting agents, and reduce the productivity of laborers (Juana et al. 2013; Farauta et al. 2011; Di Falco et al. 2011). However, people's concerns about their respective occupations are the outcomes of their observed experiences. In this study, the experiences of the respondents over the last 5 years (recall period) were collected using different sets of questions designed with a five-point Likert-type scale, which ranges from 1 (to no extent) to 5 (to a very great extent).

5.2.1 Risk Perception in Farming

The incidences of environmental change in the cultivation of Sundarban are manifested in uncertainty about the onset of the season, the effect of extreme climatic events and climatic processes, and increases in farming problems. A total of 17 distinct perceptions of climate change consequences were collected from the respondents (Table 5.3). The index score for risk perception in farming is 2.95, which is significantly different ($p = 0.046$) from the test value, $\mu = 3.0$. However, the score is not far below the average level of concern ($\mu = 3.0$) and the difference is only 0.052. The t value is -1.999, which is just at the margin of the critical value of

Table 5.3 Environmental risk perception for farming in Sundarban

Do you perceive any change in the following phenomena/incidences in Sundarban during the last 5 years?						
	Incidences	Mean ($\bar{x}$)	Standard error of mean *(SEM)*	Test value: $\mu = 3$		
				Mean difference ($\bar{x} - \mu$)	*t*	*p (two-tailed)*
A. Uncertainty in the onset of farming season						
1	Delay in onset of monsoon	2.50	0.059	−0.502	−8.576	0.000
2	Unusual early monsoon followed by weeks of dryness	2.75	0.047	−0.255	−5.477	0.000
3	Erratic rainfall in monsoon season	2.75	0.058	−0.255	−4.335	0.000
4	Heavy and long period of rainfall	3.12	0.058	0.115	1.987	0.048
5	Less rainfall in monsoon	2.18	0.062	−0.820	−13.274	0.000
6	Long period of dry season	2.79	0.051	−0.210	−4.131	0.000
Average (1–6)		**2.68**	**0.034**	**−0.321**	**−9.393**	**0.000**
B. Effect of climatic events and processes						
7	High temperature in summer	3.90	0.051	0.898	17.734	0.000
8	Severity in winter	3.40	0.057	0.400	7.008	0.000
9	Long spell of fog/smog	2.74	0.057	−0.263	−4.637	0.000
10	Overflowing of streams/rivers (flood)	2.76	0.067	−0.243	−3.612	0.000
11	Increasing risk of cyclone (number/ intensity)	2.95	0.064	−0.050	−0.777	0.438
Average (7–11)		**3.15**	**0.038**	**0.149**	**3.927**	**0.000**
C. Increase in farming problem						
12	Declining yield rate, rice (kg/bigha)	2.86	0.058	−0.145	−2.512	0.012
13	High rate of pest incidence in farming	3.29	0.061	0.290	4.793	0.000
14	Increasing diseases in farming	2.97	0.067	−0.030	−0.449	0.654
15	Rotting of rice (Aman) sapling in monsoon	2.65	0.059	−0.350	−5.946	0.000
16	Loss of soil fertility	3.20	0.053	0.203	3.811	0.000
17	Increasing soil salinity	3.33	0.060	0.325	5.383	0.000
Average (12–17)		**3.05**	**0.038**	**0.049**	**1.292**	**0.197**
Index (Average 1–17)		**2.95**	**0.026**	**−0.052**	**−1.999**	**0.046**

Scale ranges from 1 (to no extent) to 5 (to a very great extent); $n = 400$, $df = 399$, $p < 0.05$- statistically significant

Source: Computed from the field survey data

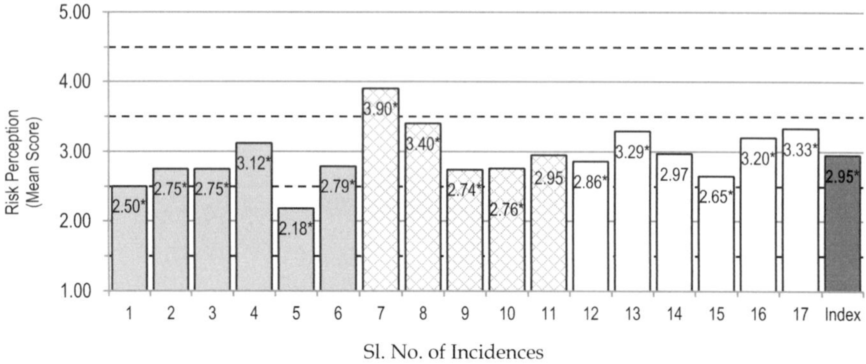

Fig. 5.4 Environmental risk perception for farming in Sundarban
Note: *Statistically significant ($p < 0.05$) difference from $\mu = 3.0$
(Source: Computed from field survey data)

the *t*-test, which ranges from ±1.96 to rounded-off ±2 for 95% *CI*. Therefore, the score can be treated as normal. The people in Sundarban perceive a moderate risk to farming in the current climatic condition. Six incidences (Fig. 5.4)—heavy and long periods of rainfall ($\bar{x} = 3.12$), high temperatures in the summer ($\bar{x} = 3.90$), severity in winter ($\bar{x} = 3.40$), a high rate of pest incidence in farming ($\bar{x} = 3.29$), loss of soil fertility ($\bar{x} = 3.20$), and increase of soil salinity ($\bar{x} = 3.33$)—are significant causes of concern in farming.

During 2001–08, monsoon rainfall significantly increased in Sundarban, at the rate of 0.0041 mm per hour (Hazra et al. 2010). Heavy rainfall with a long spell at the beginning of farming hampers the preparation of wet-beds and sowing the saplings; likewise, after the transplantation of saplings, they are rotted. The very poor sewage system in Sundarban worsens the degree of these effects. In addition, erratic rainfall in the monsoon and long dry seasons in a year affect the onset of cultivation to some extent. The surface water temperature (SST) in Sundarban has been increasing at the rate of 0.04°–0.05 °C per year (Hazra et al. 2010; Mitra et al. 2009; Khan et al. 2000). Although the pace is very slow, it is higher than the global average and even greater than the Indian Ocean (IPCC 2007). The air temperature has been increasing at a more alarming rate: 0.1058 °C per year (or 1 °C per decade) during 2002–09 (Hazra et al. 2002, 2010). These concerns are prominent in the cognitive minds of the respondents.

The temperature increase in summer and severity in winter affects the farming in Sundarban to a great extent. Beyond a certain range of temperature, the yield rate of the crops is reduced. Every 2 °C increase in mean air temperature is estimated to decrease the rice yield by 0.06 tons per hectare in the coastal areas (Bhattacharya 2007). A very low yield rate of *Aman* rice in Sundarban (2391 kg./hectare in 2012–13) compared to the state average (2596 kg./hectare) may be the probable consequence of that. In addition, high temperatures influence the ability of plants to obtain and use the moisture, and accelerates transpiration and soil evaporation. In Sundarban, sources of irrigation (ponds and canals) are limited. An increase in temperature causes extreme water shortages in summer. Therefore, practicing *Rabi* cultivation is risky if farmers do not have plenty of water storage. Moreover, a warm-humid summer and cold-dry winter are salient climatic characteristics of Sundarban. In extreme weather conditions, farmers are easily exhausted, which reduces labor productivity. Extreme climatic events, such as tropical cyclonic storms, immensely affect farming. The frequency and intensity of SCS have significantly increased over the Bay of Bengal during intense cyclonic months (May, October, and November) in the past 100 years (Singh 2007; Singh et al. 2001; Khan et al. 2000).

Soil salinization is another great concern of farmers in Sundarban. Salinity beyond 6.0 ppt hampers rice cultivation. In Sundarban, it exceeds the permissible limit (Ghosh 2012; Goshthakur 2009). Moreover, higher salinity is one of the leading causes of declining soil fertility in Sundarban. Apart from soil salinity and fertility, increasing pest incidence and disease are moderate concerns among the farmers.

The mean score for migrants (Table 5.4) in risk perception for farming is 2.883 which is significantly ($t = 3.017$, $p = 0.003$) lower than the normal level of concern ($\mu = 3$). However, non-migrants ($\bar{x} = 3.012$, $t = 0.356$, $p = 0.722$) perceive moderate risk in farming. There is a statistically significant difference ($t = -2.486$, $df = 398$, $p = 0.015$) between the groups in terms of risk perception. The slightly higher (0.129) risk perception among the non-migrants compared with migrants may be attributed to an intense attachment to the origin by the occupation. It is also an indication of the potential for migration from Sundarban.

5.2.2 Risk Perception in Fishing

Fishing in Sundarban includes the collection of both fish and non-timber forest products. The fishers' concern in Sundarban can be attributed to two categories: worsening of the collecting environment and effects on production. The mean score

Table 5.4 Risk perception for farming and migration in Sundarban (Independent sample *t*-test)

Statistics	Index (Risk in farming)	
	Migrants	Non-migrants
Mean	2.883	3.012
SEM	0.039	0.035
N	200	200
Mean difference *(Mig. – Non-mig.)*	−0.129	
t (independent sample)	−2.486	
p (two-tailed)	0.013	
Eta Squared (effect size)	0.015	
Df	398	

Source: Computed from field survey data

($\bar{x}$) for risk perception in fishing is 2.96, which is not statistically significantly ($p = 0.246$) different from $\mu = 3.0$ (Table 5.5). The mean difference of ($\bar{x}$-μ) is −0.040. Therefore, Sundarban people perceive a moderate level of risk in fishing.

The respondents are highly concerned about declines in fish production ($\bar{x} = 3.76$, $p = 0.000$) (Fig. 5.5). Reductions in fish stocks and varieties are also concerns ($\bar{x} = 3.49$, $p = 0.000$), as are increases in trash fish[1] ($\bar{x} = 3.34$, $p = 0.000$) and declines in both the quality and quantity of production. In 2009, the Department of Sundarban Affairs reported that fishing efforts had doubled during the past 15 years, but Catch per Unit Effort (CPUE)[2] drastically declined to 58–65 kg. per haul from 150 to 200 kg. per haul. During 1995 to 2005, tiger shrimp was abundantly available in rivers and creeks and was easily caught by a drawing net with the finest weave. The voracious extraction of the shrimp in an unsustainable manner has made it into a short-lived activity (Feature Box 3.1).

Moreover, environmental legislations are one of the great barriers to fishing in Sundarban (Mistri and Das 2018). This phenomenon has resonated in the concerns of fishermen. The mean score for environmental legislation indicates a great extent of risk perception ($\bar{x} = 3.18$, $p = 0.024$). As with farming, extreme weather events—especially the increased risk of cyclonic storms—threaten fishing. Fishers perceive a moderate extent of risk ($\bar{x}$=2.94, $p = 0.384$) from these events. The fishers often catch for at least a week at a stretch. However, in severe cyclone-prone months, they often have to leave the fishing ground before the expected schedule. Every year, some fishers go missing after gusty winds or cyclonic storms. In addition, fishing in Sundarban is affected by riverine conditions. The decline of river and creek depths ($\bar{x} = 2.92$, $p = 0.224$) and an increase in tidal force or currents ($\bar{x} = 2.92$, $p = 0.217$) are also concerns of fishermen (to a moderate extent) in Sundarban. Shallow river channels are not considered to be suitable fishing ground. In shallow rivers and creeks, water passes through a particular channel at the time of low tide, resulting in increased tidal force. Moreover, sand bars hamper the free movement of fish and the seasonal migration of some species. Both the migrants ($\bar{x} = 2.916$, $t = -1.810$, $p = 0.073$) and non-migrants ($\bar{x} = 3.003$, $t = 0.068$, $p = 0.946$) perceive more or less equal extent of (moderate) risk in fishing, with no statistically significant difference between the groups (Table 5.6).

Finally, it can be concluded that the people of Sundarban are not exempted from the ill effects of the changing environment. They are moderately concerned about the environmental changes that can affect their livelihoods. The people are more concerned about micro-risks to their own families, communities, and local incidences rather than holistic concerns about Sundarban as a whole, geographically distant people and places, or non-human nature such as wildlife and mangroves. Cyclones, water surge, salinization, rainfall, and temperature change are greater causes of concern in their cognitive minds. Both migrants and non-migrants perceive more or less equal environmental risk in their respective activities; no significant differences were observed between the groups in terms of the same.

[1]Trash fish or rough fish are less desirable to the catchers in a region. Environmentalists suggest that the production of rough fish is increased due to climate change.

[2]Catch per Unit Effort (CPUE) refers to total amount of fish that are caught by a boat in kg./haul.

Table 5.5 Environmental risk perception for fishing in Sundarban

Do you perceive a change in the following phenomena/incidences in Sundarban during the last 5 years?

	Incidences	Mean ($\bar{x}$)	Standard error of mean *(SEM)*	Test value: $\mu = 3$ Mean difference ($\bar{x} - \mu$)	*t*	*p (2-tailed)*
A. Worsening of the collecting environment						
1	Increasing risk of cyclone/low pressure/gusty winds	2.94	0.068	−0.060	−0.872	0.384
2	Increasing risk of flash floods/high tide	2.84	0.067	−0.160	−2.399	0.017
3	Increasing risk of tidal force/river or ocean current	2.92	0.063	−0.078	−1.237	0.217
4	High humidity and temperatures in summer	2.62	0.066	−0.384	−5.789	0.000
5	Severe winter (windy and low temperature)	2.81	0.061	−0.194	−3.171	0.002
6	Declining depth of rivers and creeks	2.92	0.064	−0.078	−1.220	0.224
Average (1–6)		**2.84**	**0.039**	**−0.159**	**−4.053**	**0.000**
B. Effect on production						
7	Declining fish (fish/prawn/crab) production	3.76	0.084	0.761	9.046	0.000
8	Reduction of fish/prawn/crab stocks and varieties	3.49	0.088	0.485	5.488	0.000
9	Increasing trash fish	3.34	0.079	0.343	4.327	0.000
10	Faster spoilage of the catch	1.96	0.067	−1.041	−15.554	0.000
11	Increasing salinity of river/ocean water	2.95	0.062	−0.052	−0.842	0.400
12	Decrease in the water transparency (or green color)	2.76	0.062	−0.239	−3.868	0.000
13	Environmental regulations affecting fishing	3.18	0.081	0.183	2.270	0.024
Average (7–13)		**3.06**	**0.041**	**0.063**	**1.521**	**0.130**
Index (1–13)		**2.96**	**0.034**	**−0.040**	**−1.162**	**0.246**

Scale ranges from 1 (to no extent) to 5 (to a very great extent), $n = 268$, $df = 267$, $p < 0.05$-statistically significant

Source: Computed from the field survey data

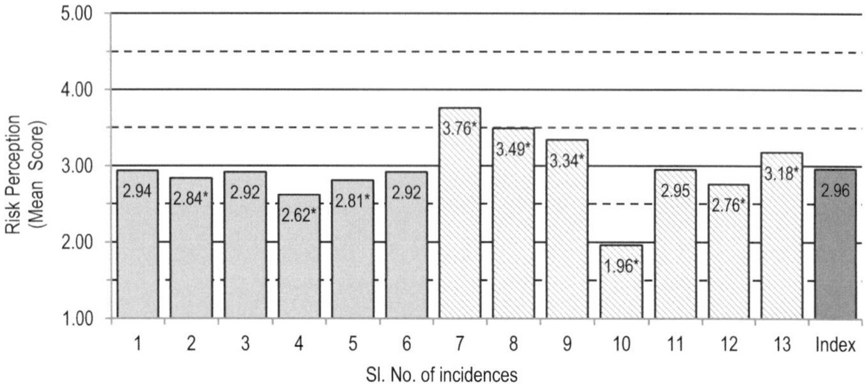

Fig. 5.5 Environmental risk perception for fishing in Sundarban
Note: *Statistically significant ($p < 0.05$) difference from $\mu = 3.0$
(Source: Computed from field survey data)

Table 5.6 Risk perception for fishing and migration in Sundarban (Independent sample *t*-test)

Statistics	Index (Risk in fishing)	
	Migrants	Non-migrants
Mean	2.916	3.003
SEM	0.046	0.050
N	132	136
Mean difference	−0.087	
(Mig. – Non-mig.)		
t (independent sample)	−1.282	
p (two-tailed)	0.201	
Eta Squared (effect size)	0.006	
Df	266	

Source: Computed from field survey data

References

ADB. (2009). *Climate change and migration in Asia and Pacific: Executive summery*. Philippines: Asian Development Bank (ADB).

Bhattacharya, S. (2007). *Lesson learn for vulnerability and adaptation assessment from India's first national communication*. Paper 7 of BASIC project, Winrock International. http://www.basic-project.net/data/final/Paper07India%20Lessons%20for%20VA%20from%20First%20National%20Communicat%85.pdf. Accessed 30 Sept 2016.

Chand, B. K., Trivedi, R. K., Dubey, S. K., & Beg, M. M. (2012). *Aquaculture in changing climate of Sundarban: Survey report on climate change vulnerabilities, aquaculture practices and coping measures in Sagar and Basanti Blocks of Indian Sundarban*. Kolkata: West Bengal University of Animal and Fishery Sciences.

Di Falco, S., Yesuf, M., Kohlin, G., & Ringler, C. (2011). Estimating the impact of climate change on agriculture in low-income countries: Household level evidence from the Nile Basin, Ethiopia. *Environmental and Resource Economics, 52*(4), 457–478.

Edvardsson, I. R., Tingley, D., Conides, A. J., Drakeford, B., & Holm, D. (2011). Fishermen's risk perception in four European countries. *MAST, 10*(1), 139–159.

Farauta, B. K., Egbule, C. L., Idrisa, Y. L., & Agu, V. C. (2011). *Farmers' perceptions of climate change and adaptation strategies in Northern Nigeria: An empirical assessment.* Nairobi: African Technology Policy Studies Network.

Ghosh, A. (2012). *Living with changing climate: Impact, vulnerability and adoption challenges in Indian Sundarbans.* New Delhi: Centre for Science and Environment (CSE). http://cseindia.org/userfiles/Living%20with%20changing%20climate%20report%20low%20res.pdf. Accessed 20 Dec 2013.

Goshthakur, D. (2009, December 4). Sundarbaner Bastutantra Nonarbish (Poison of salinity in the ecosystem of Sundarban). *Annadabazar Patrika.* p. 4.

Hazra, S., Ghosh, T., Dasgupta, R., & Sen, G. (2002). Sea level and associated changes in the Sundarbans. *Science and Culture, 68*(9–12), 309–321.

Hazra, S., Samanta, K., Mukhopadhyay, A., & Akhandet, A. (2010). *Temporal change detection (2001–2008) study of Sundarban.* Final report. School of Oceanographic Studies, Jadavpur University. http://www.iczmpwb.org/main/pdf/ebooks/WWF_FinalReportPDF.pdf. Accessed 23 Jan 2013.

IOM. (2009). *Migration, environment and climate change: Assessing the evidence.* In: F. Laczko & C. Aghazarm (Eds.). Geneva: IOM.

IPCC. (2007). *Climate change 2007: The physical science basis* (Contribution of working group I to the fourth assessment report of the intergovernmental panel on climate change). Cambridge: Cambridge University Press.

Juana, J. S., Kahaka, Z., & Okurut, F. N. (2013). Farmers' perceptions and adaptations to climate change in Sub-Sahara Africa: A synthesis of empirical studies and implications for public policy in african agriculture. *Journal of Agricultural Science, 5*(4), 121–135.

Khan, T. M. A., Singh, O., & Rahaman, M. S. (2000). Recent sea level and sea surface temperature trends along the Bangladesh coast in relation to the frequency of intense cyclone. *Marine Geodesy, 23*(2), 23–103.

Kniveton, D., Schmidt-Verkerk, K., Smith, C., & Black, R. (2008). *Climate change and migration: Improving methodologies to estimate flows.* Geneva: IOM.

Leiserowitz, A. A. (2005). American risk perceptions: Is climate change dangerous? *Risk Analysis, 25*(6), 1433–1442.

Leiserowitz, A. A. (2006). Climate change risk perception and policy preferences: The role of affect, imagery, and values. *Climatic Change, 77*(2), 45–72.

Mistri, A., & Das, B. (2018). Migration in response to environmental change: A risk perception study from Sundarban biosphere reserve. In S. Irudaya Rajan & R. B. Bhagat (Eds.), *Climate change, vulnerability and migration.* New Delhi: Routledge.

Mitra, A., Gangopadhyay, A., Dube, A., Schmidt, A. C. K., & Banerjee, K. (2009). Observed changes in water mass properties in the Indian Sundarbans (Northwestern Bay of Bengal) during 1980–2007. *Current Science, 97*(10), 1445–1452.

Myers, N. (2005). Environmental refugees: An emergent security issue. In *13th economic forum*, 23–27 May 2005, Prague.

O'conner, R. E., Bord, R. J., & Fisher, A. (1999). Risk perceptions, general environmental beliefs, and willingness to address climate change. *Risk Analysis, 19*(3), 461–471.

Singh, O. P. (2007). Long-term trends in the frequency of severe cyclones of bay of Bengal: Observations and simulations. *Mausam, 58*(1), 59–66.

Singh, O. P., Khan, T. M. A., & Rahman, M. S. (2001). Has the frequency of intense tropical cyclones increased in the North Indian Ocean? *Current Science, 80*(4), 575–580.

Sjoberg, L. (2000). The methodology of risk perception research. *Quality and Quantity, 34*(4), 407–418.

Slovic, P. (1987). Perception of risk. *Science, 236*(4799), 280–285.

Slovic, P., & Peters, E. (2006). Risk perception and affect. *Current Directions in Psychological Science, 15*(6), 322–325.

Livelihood Assets and Migration

6

Abstract

The asset possession of the household immensely influences the decision of migration. Livelihood assets are classified into five categories, namely, human, financial, natural, physical and social capital. These are not mutually exclusive, and one compensates to other for the deterioration of the same. Assets endow the people to cope with the adverse situation and help to explore diverse livelihood options for getting out of risk. This chapter looks insight into the detailed investigation of the above-mentioned asset possession and compares between migrants and non-migrants in terms of the same.

Keywords

Livelihood assets · Human capital · Social network · Physical capital · Natural capital

Migration is one of the livelihood strategies. A livelihood consists of the capabilities, assets and activities required for a means of living (Camber and Conway 1991). Assets or capital endowment of people implies people's strengths to cope with the adverse situation and that ultimately influence the decision of migration. People require a range of assets, such as human, financial, physical, natural and social capital, to achieve positive livelihood outcomes (DFID 2001). A single category of assets is not sufficient to yield all the many and diverse livelihood outcomes that people seek for descent living. The present chapter gives a broad view of the above-mentioned five categories of assets, and compares the same between the exposed group (migrants) and the control group (non-migrants). This chapter comprises four sections. Both the primary and secondary data sources are taken help for the discussion. Various descriptive and inferential statistics, such as *t*-test, x^2 test and composite indices, are incorporated to delineate the data sets.

A. Mistri, B. Das, *Environmental Change, Livelihood Issues and Migration*,
https://doi.org/10.1007/978-981-13-8735-7_6

6.1 Human Capital

Human capital is the intrinsic virtue of human being like education and knowledge, skill, ability to labour and good health. These virtues together enable people to pursue different strategies and achieve their livelihood objectives. Moreover, endowment with human capital makes an ability to convert other assets into positive livelihood outcomes or use it into systematic ways. At a household level, human capital is a factor of the amount and quality of labour available, which varies according to household size, skill levels, leadership potential, health status, etc. (DFID 2001).

6.1.1 Education

Education is the most precious asset of a human being, and it is completely his/her own. In general, the literate person gets more advantage in migration compared to the illiterate one. Education reduces the psychic cost as well as the opportunity cost of migration. The distance covered through migration is significantly related with the level of education. Moreover, all sorts of rational consideration (thesis), such as cost benefit (Todaro 1969; Sjaastad 1962), self-enforcement and altruism (Stark and Lucas 1988), dissatisfaction (Spear 1974) and relative deprivation (Stark 1984), threshold of a stress (Wolpert 1966) and adaptive response (McLeman and Smith 2006), are highly advocated by different levels of education of the migrants.

Nearly 80% migrants in the study are literate, which includes 41.5% middle and secondary education, 25.5% primary education and 8.0% higher secondary and above (Table 6.1). Likewise, most of the non-migrants (76.6%) are literate. The state (West Bengal) average literacy rate is 77.08%, and national average is 74.04% (Census of India 2011). The share of illiteracy among the non-migrants is slightly higher compared to the migrants, which are 27% and 21%, respectively. However, a chi-square test for independence (with Yates Continuity Correction) reveals that there is no significant association between the status of migration and level of

Table 6.1 Educational status of respondents in the Sundarban

	Migrants		Non-migrants		Total	
Level of educational	Frequency	Percent	Frequency	Percent	Frequency	Percent
Illiterate	41	20.5	53	26.5	94	23.5
Up to primary (till class V)	51	25.5	53	26.5	104	26.0
Middle and secondary (class X)	83	41.5	64	32.0	147	36.8
Higher secondary and above	16	8.0	27	13.5	43	10.8
Other	9	4.5	3	1.5	12	3.0
Total	200	100	200	100	400	100

Source: Field survey data

education. The x^2 value is 1.68 ($df = 1$, $n = 400$), $p = 0.19$, and strength of association (effect size) between the variables (Feature Box 6.1) is weak ($phi = -0.07$).

Feature Box 6.1: Effect Size for *t*-Test and Chi-Square (x^2) Test

The probability value (p) does not infer the degree to which the two variables are associated with one another. In case of large sample, very small difference between two groups becomes statistically significant. Effect size is a set of statistic that indicates the relative magnitude of the differences between means or the amount of total variance in the dependent variable is explained by the predictor or independent variable. It is also known as strength of association. The most commonly used techniques for the effect size are eta squared and Cohen's d.

In *t*-test, eta squared is commonly used to estimate the effect size. The formula is

$$\frac{t2}{t2+(n1+n2-2)}$$

The value of eta squared ranges from 0 to 1.0. The guidelines proposed by Cohen (1988) for interpretation of the eta squared value are 0.01, small effect; 0.06, moderate effect; and 0.14, large effect.

In chi-square (x^2) test, for 2 (column) by 2 (row) crosstabs, phi coefficient is followed to examine the effect size between two variables. Cohen's criteria for phi coefficient are 0.10, small effect; 0.30, medium effect; and 0.50, large effect. The crosstabs with more than 2 by 2, Cramer's V criterion is followed to examine the effect size; slightly different criteria are recommended for judging the size of the effect. To determine which criteria to use, first subtract 1 from the number of categories in your row variable (R–1) and then subtract 1 from the number of categories in your column variable (C–1). Pick whichever of these values is smaller. For R–1 or C–1 equal to 1 (two categories): small = 0.01, medium = 0.30, and large = 0.50. For either R–1 or C–1 equal to 2 (three categories): small = 0.07, medium = 0.21, and large = 0.35. For either R–1 or C–1 equal to 3 (four categories): small = 0.06, medium = 0.17, and large = 0.29.

6.1.2 Household Size

Household size indicates the family burden on earning member. Studies suggest (Massey 1990; Stark and Lucas 1988; Spear 1974) that large family diverse the labour portfolio through migration. In the rural areas of the developing countries, more siblings put the responsibility on the elder, and he/she is obligated to migrate

Table 6.2 Household size in the Sundarban

Statistics	Total HH	Migrated HH	Non-migrated HH
Mean	4.66	4.81	4.51
Mode	4.0	4.0	4.0
Std. deviation	1.80	1.80	1.79
Minimum	1.0	1.0	1.0
Maximum	12.0	11.0	12.0
Observation	400.0	200.0	200.0

Source: Field survey data

for the betterment of his/her family. It is denoted as 'self-enforcing' migration (Stark and Lucas 1988). In the twenty-first century, most of the society begins to believe in nuclear family. They are more concerned about child rearing and their future. The parents who have higher altruism towards children are more likely to migrate (Stark and Lucas 1988). The mean household size[1] of the study is 4.7; household size for migrants and non-migrants is 4.8 and 4.5, respectively (Table 6.2). The modal class for household size is 4.0; there is no variation in household size between the groups.

Until now, in a patriarchal society like India, men are main bread earner in the family. Higher sex ratio at household level, therefore, often creates pressure on the male earners. In the study, the sex ratio[2] among migrated households (967 female per 1000 male) is fairly higher than the non-migrated households (952 female per 1000 male). It infers a push which is emerged from household level to move out of the Sundarban.

6.1.3 Age Structure

More members in a family do not only mean more mouths to consume but also mean more hands to work. Youths are more likely to migrate. They are ready to take challenges at unknown destinations. Now, India is at the pick of demography dividend. The Sundarban is not an exception. More than two-third (68%) of the people are reported in working age (Table 6.3), which is far higher than the national average (60.5%) as well as the state average (64.4%) (Census of India 2011).

6.1.4 Health Status

Healthy people not only able to take challenges to a unknown or unfamiliar environment at destination and fit for any sorts of physical labour work but also increase the production; less amount is drained out for ailment vis-à-vis more

[1]In 2011 Census, the mean household size in India is 4.9 and West Bengal is 4.6.

[2]Overall sex ratio in India is 940 female per 1000 male, and in West Bengal, it is 947 female per 1000 male in 2011 Census.

Table 6.3 Age structure in the Sundarban

	Migrated HH		Non-migrated HH		Total HH	
Age group (Yrs.)	Frequency	Percent	Frequency	Percent	Frequency	Percent
0–14	223	23.3	211	23.4	434	23.4
15–59	654	68.3	612	68.0	1266	68.1
60+	81	8.5	77	8.6	158	8.5
Total	958	100	900	100	1858	100

Source: Field survey data

amount for saving and investment. The health of migrant labourer is the growing concern in the contemporary India. Migrants suffer lots of health hazards and carry communicable diseases from one place to another. They live in no and minimal availability of basic amenities including safe drinking water and sanitation at the worksites. The Interstate Migrant Workmen (Regulation of Employment and Condition of Services) Act of 1979 and The Unorganised Workers' Social Security Act of 2008 were introduced to ensure the provision of health care, such as suitable residential accommodation and sanitation facilities, medical facilities, protective clothing and other safety net and social security like health and maternity benefits, life and disability cover and old age protection of the migrant labourers. But these are not implemented properly by the central government as well as the state governments. Nearly 200 million person-days of work are wasted in India, and the country loses about Rs. 366 billion each year due to the ill effects of polluted water and inadequate sanitation of facilities (IDFC 2011, p.228). The total amount of estimated expenditure for its related diseases is nearly Rs. 470–610 billion per annum (World Bank 2013, p. 16).

In the study, 68% respondents (both the migrants and migrants) reported that they did not suffer from any health problem or diseases during the last 365 days (Table 6.4). On the other hand, around 32% migrants had suffered from ill health during that period. Due to ill health, 47% migrants wasted in between 30 and 90 working days in a year; it is followed by 38% wasted more than 90 working days and 16% wasted less than 30 working days. There is not much variation at the cost of ailment between the migrant and non-migrant. Expenditure for the ailment of the sick person widely varies from a minimum of Rs. 100 to a maximum of Rs. 250,000. Therefore, the modal expenditure[3] is Rs. 5000 that means the highest proportion of the people spent Rs. 5000 for their ailment. Apart from the point estimation, nearly 47% sick people spent Rs. 5,000 and below. It is followed by 28.3% spent in between Rs. 5,000 and 15,000, and the rest (25.2%) spent around Rs. 15,000 and above (Table 6.5).

Overall, islanders are full of basic human capitals. Most of them are young, educated and possess good health. Hence, they are very energetic to migrate in long distance, such as interstate, and circulate over the year. A least number of working

[3]Mean expenditure is violated by the outlier. The mean health expenditure is Rs 15,443.

Table 6.4 Work affected by ill health during the last 365 days

		Migrants		Non-migrants		Total	
Work affected by ill health		Frequency	Percent	Frequency	Percent	Frequency	Percent
	Yes	64	32.0	63	31.5	127	31.8
	No	136	68.0	137	68.5	273	68.3
	Total	200	100	200	100	400	100

Source: Field survey data

Table 6.5 Amount (Rs.) spends for ailment

Amount (Rs.)	Total (Migrants + non-migrants)	Percent
Rs. 5,000 and below	59	46.5
Rs. 5000–15,000	36	28.3
Rs. 15,000 and above	32	25.2
Total	127	100.0

Source: Field survey data

days are wasted due to illness. Meanwhile, a very few amount is drained out for the same. It may facilitate good amount of remitting regularly into the Sundarban.

6.2 Financial Capital

Financial capital, in the lexicon of the economics, is the money, credit and other forms of stock or funding that individual or company use to spend for consumption or invest for production. It has a monetary or exchange value which is the key distinguished character in relation to other types of capitals. Financial capital in terms of flows, such as income, remittances, pension, or other transfer of states or government, etc., and stocks like cash, bank deposits, or liquid assets, such as livestock, jewellery and other durable assets, influence the decision of migration or retention at the origin.

6.2.1 Durable Assets Possession

Durable assets are stocks of resources that are accumulated and hold over the time for future consumption and security against contingencies. Moreover, the asset poverty approach is considered as one of the important poverty measures compared to income approaches because poverty is not necessarily only low income but also includes a low wealth possession. Apart from the direct economic return, assets have immense indirect return also. For an example, possession of a mobile phone enhances social network of labourer, helps to easily access the information regarding the job that reduces the opportunity cost, such as lessening waiting time for getting a job, to be absorbed in a permanent and high paid job, information regarding

Table 6.6 Durable asset possession by the households in the Sundarban

Sl. No.	Durable assets	Migrated HH		Non-migrated HH		Total HH	
		Frequency	Percent	Frequency	Percent	Frequency	Percent
1	Radio/ transistor	44	22.0	52	26.0	96	24.0
2	Television (TV)	59	29.5	58	29.0	117	29.3
3	Telephone (mobile/ landline/ both)	160	80.0	151	75.5	311	77.8
4	Bicycle/ paddle van	95	47.5	88	44.0	183	45.8
5	Scooter/ motorcycle/ moped	6	3.0	11	5.5	17	4.3
6	Car/jeep/ motor van/tractor/ power tiller/ other four wheelers	1	0.5	1	0.5	2	0.5
7	Computer/ laptop	3	1.5	0	0.0	3	0.8
8	Machinery (generator, water pump sets, spray and rice threshing and others)	23	11.5	21	10.5	44	11.0
9	Fishing contrivances (boat/craft, BLCs, net and others)	44	22.0	77	38.5	121	30.3
10	None of the specified assets possession	24	12.0	22	11.0	46	11.5

Source: Field survey data

transportation and accommodation and so on. The accounts of some basic asset possession by the households in the Sundarban are portrayed here (Table 6.6). Only one asset, telephone, especially the mobile phone, is possessed by most of the households (78%) in the Sundarban. The other major assets are bicycle or paddle van (46%), fishing contrivances (30%) and television (29%). Comparatively, costly assets like two wheeler, four wheeler, computer or laptop are hardly possessed by the households. On the other hand, around 12% households possess none of the above specified assets. With the help of all asset possession (excluding two and four wheeler and computer or laptop), a durable asset index (DAI) is constructed to compare between migrated and non-migrated households as a whole.

The DAI is prepared with the help of principal component analysis (PCA) with varimax rotation method. The three-component solution explains a total of 63.3% of the variance. An average of the three-component scores is used to construct durable asset index (DAI) for the individual household. All variables are unidirectional and comprised of two categories, viz., *yes-1* for having a specific asset and *no-2* for not having the same. Therefore, households with higher DAI score infer the lower asset possession. The formula is as follows:

$$C_1 = W_1 * X_1 + W_2 * X_2 + W_3 * W_3 \ldots\ldots\ldots\ldots\ldots + W_6 * X_6$$

where:

C = Score of the component

X = Standardised variable (individual asset possession)

W = Component score coefficient (weightage)

The scores of the index vary from a minimum of −1.62 to a maximum of 1.02 (Table 6.7). Most of the households (64%) score in between −0.5 and 0.50. However, an independent sample *t*-test ($t = 1.04$, $df = 398$) reveals no significant difference ($p = 0.30$) in the mean (DAI) scores between the migrated ($M = 0.03$, $SD = 0.575$) and non-migrated ($M = -0.03$, $SD = 0.579$) households. The effect size (eta squared = 0.003) is very low. It infers that only 0.30% of the variance in the DAI is explained by the status of migration. The people in the Sundarban irrespective of migration status possesses very poor amount of durable assets.

Table 6.7 Durable asset index (DAI)

Statistics	Migrated HH	Non-migrated HH	Total HH
Mean	0.03	−0.03	0.00
Median	0.08	0.01	0.01
Mode	0.42	1.02	0.42
Std. deviation	0.58	0.58	0.58
Minimum	−1.62	−1.33	−1.62
Maximum	1.02	1.02	1.02
Skewness	−0.04	0.06	0.01
Observation	200	200	400

	Migrated HH		Non-migrated HH		Total	
DAI score	Frequency	Percent	Frequency	Percent	Frequency	Percent
−0.5 and below	37	18.5	44	22.0	81	20.3
−0.5–0.00	59	29.5	51	25.5	110	27.5
0.00–0.5	69	34.5	77	38.5	146	36.5
0.5 and above	35	17.5	28	14.0	63	15.8
Total	200	100	200	100	400	100

Source: Compute from field survey data

It is important to mention here that asset creation is dependent on the everyday requirement along with affordability of the people. In the Sundarban, most of the villages are not electrified till date, which restricts the use of television, computer or laptop and other electronic gadgets. Moreover, their economic and social utility is less because poor infrastructure cannot support its maximum utilisation. Maintenance and repairing of the valuable and sophisticated modern assets are also problematic. Poor road condition does not allow large number of vehicles. Even uses of four wheelers are restricted within the islands because transportation from one island to another is hardly accessible through vessels or bridge; small ferry boat is the only option.

6.2.2 Livestock

The livestock sector contributes around 6.8% of gross domestic product (GDP) in India and employs 8.0% of the labour force (FAO 2005, p.6). Livestock is a very important liquid asset of the rural households. Rural households often sell their domestic pets to meet the cost incurred for migration or at the time of sudden moving out. When husband migrates and unable to send money to his wife or family in proper time, the livestock provides them temporary financial support.

In the Sundarban, cow, goat, pigs and poultry, such as different kinds of chicken and duck or swan, are most common domestic pets. State governments or Tiger Reserve Authority seldom provide loans for livestock, especially for goat, pig and poultry farming. Around 73% households possess at least one of the animals mentioned above, but the number is far from satisfactory, and economically valuable domestic pets, such as cow or buffalo, pig and sheep or goat, are scanty. Cattle or buffalo, sheep or goat and pig are reared by 38%, 32%- and 9.0% households, respectively, whereas 60% households pet poultry. Due to different economic valuation and for international comparison, Food and Agriculture Organization (FAO) has designed country-specific coefficient or conversion factors for the different group of livestock units. In the study, a standard livestock unit (SLU) index is developed with the help of coefficient (weightage) designed by FAO (2005, p. 6; 2011, p. 37) for India. The formula is as follows:

$$\begin{aligned} \textbf{\textit{SLU Index}} = {} & (\mathbf{0.50}{*}\textit{\textbf{No.of Cattle or Buffalo}}) + (\mathbf{0.10}{*}\textit{\textbf{No.of Sheep or Goat}}) \\ & + (\mathbf{0.20}{*}\textit{\textbf{No.of Pig}}) + (\mathbf{0.01}{*}\textit{\textbf{No.of Poultry}}) \end{aligned}$$

The households with higher index score get more financial benefit from livestock. The scores range from 0.00 to 4.02 (Table 6.8). The mean score is 0.54 which is slightly higher for the non-migrated households (0.56). Nearly 26% households score 0.00, means do not possess any livestock unit. Around 50% households score in between 0.01 and 1.0, and the rest are above that.

Table 6.8 Standard livestock unit (SLU) index

Statistics	Migrated HH	Non-migrated HH	Total HH
Mean	0.53	0.56	0.54
Median	0.27	0.30	0.30
Mode	0.00	0.00	0.00
Std. Deviation	0.71	0.63	0.67
Minimum	0.00	0.00	0.00
Maximum	4.02	2.56	4.02
Skewness	2.08	1.14	1.68
Observation	200	200	400
	Migrated HH	Non-migrated HH	Total HH
SLU Index (Score)	Percent	Percent	Percent
1.0 & above	21.0	28.0	24.5
0.01–1.00	54.5	45.0	49.8
0.00	24.5	27.0	25.8
Total	100.0	100.0	100.0

Source: Computed from field survey data

The statistical test ($t = -0.459$, $df = 398$) unveils that there is no significant ($p = 0.65$) difference in the mean SLU score between migrated ($\bar{x} = 0.53$ $SD = 0.71$) and non-migrated households ($\bar{x} = 0.56$, $SD = 0.67$), and the magnitude of the difference (eta squared = 0.005) is very minimal.

6.2.3 Household Types

The household type is defined according to the source from which a household's 50% or more income comes from. The family income during the last 365 days has been considered as the reference period for the determination of household type. Survey from 400 households expresses that nearly 51% households, which include 24% cultivation, 9.0% agricultural labour and 18% fishing, have reported under farming and fishing household types (Table 6.9). There is a significant ($t = 104.51$, $df = 6$, $n = 400$), $p = 0.00$) and strong difference (Cramer's V = 0.51) between the migrated and non-migrated households in terms of major sources of income or household types. Around 70% of the non-migrated households depend on the agriculture sector, whereas most of the migrated families, around 68%, depend on non-farming sector. Within the non-farming sector, daily labour or casual work (46%) is predominant means of living compared to others. It has already been discussed in the section, 'work engagement of migrants: before and after move' in Chap. 5 that more than half of migrants are engaged in the construction sector as a daily or causal wage labourer at the place of destinations. Migrants from the Sundarban are basically unskilled or semi-skilled as they are not highly educated.

Most of them are (67%) educated up to primary and secondary classes, and even 21% are illiterate. They used to work as farmers, agro-labourers and fishermen or

Table 6.9 Household types in the Sundarban

	Migrated HH		Non-migrated HH		Total HH	
HH Types	Frequency	Percent	Frequency	Percent	Frequency	Percent
Cultivation	32	16.0	65	32.5	97	24.3
Agricultural labour	13	6.5	21	10.5	34	8.5
Fishing and NTFPs	19	9.5	53	26.5	72	18.0
Daily labour/casual labour	92	46.0	10	5.0	102	25.5
Regular wage/salary earning	21	10.5	17	8.5	38	9.5
Business	8	4.0	24	12.0	32	8.0
Others (including HHI and T and C)	15	7.5	10	5.0	25	6.3
Total	200	100	200	100	400	100

Note: HHI- Household Industry, T&C- Transport and Communication
Source: Field survey data

Table 6.10 Descriptive statistics related to household income in the Sundarban

Statistics	Total HH (Rs./month)	Migrated HH (Rs./month)	Non-migrated HH (Rs./month)
Average	4,624	4,948	4,301
Median	3,000	3,500	3,000
Mode	3,000	3,000	3,000
Std. Deviation	5,042	4,838	5,230
Minimum	400	1,000	400
Maximum	50,000	50,000	40,000
Skewness	4.56	0.17	0.17
Observation	400	200	200

	Migrated HH		Non-migrated HH		Total	
Income (Rs./month)	Frequency	Percent	Frequency	Percent	Frequency	Percent
Rs.10,000 and above	17	8.5	15	7.5	32	8.0
Rs.3000–10,000	87	43.5	54	27.0	141	35.3
Rs.3000 or below	96	48.0	131	65.5	227	56.8
Total	200	100	200	100	400	100

Source: Computed from field survey data

daily labourers. Hence, their scope of engagement is limited to the unorganised sectors, mostly in construction sector, at the destinations.

6.2.4 Family Income

The monthly income of the households widely varies from a minimum of Rs. 400 to a maximum of Rs. 50,000 (Table 6.10). Average income, therefore, is affected by the outlier and produces a skewed distribution of the same. The median household income, a half way or point of income distribution, gives a more accurate representation of average monthly income of the family. The median monthly income[4] of migrated households (Rs. 3500) is greater than non-migrated households (Rs. 3000). The minimum monthly income of migrated households begins with Rs. 1000, where minimum monthly income of non-migrated household is around Rs. 400 which is two times lower than the former. Nearly 48% migrated households earn Rs. 3000 and below, 44% earn in between Rs. 3000 and Rs. 10,000 and the rest (nearly 9.0%) earn Rs. 10,000 and above. On the other than, nearly 66% non-migrated households earn Rs. 3000 and below, around 27% households earn in between Rs. 3000 and Rs. 10,000 and the rest (8.0%) earn Rs. 10,000 and above.

The statistical test also ($x^2 = 13.25$, $df = 2$, $n = 400$) suggests that there is a significant association ($p = 0.001$) between monthly family income and migration status (migrants or non-migrants) of the households.

6.3 Natural Capital

Natural capital is 'the natural resource stocks from which resource flows and services useful for livelihoods are derived' (DFID 2001, p. 23). The natural capitals which provide the people free goods and services are often called as ecosystem services. The natural resource comprises wide varieties of resources from 'intangible public goods', such as atmosphere, ecotourism and biodiversity to 'divisible assets' used directly for production like land, forest and wild resources, water and marine resources, etc. In the Sundarban, natural capital plays a crucial role because most of the people derive entire or part of their livelihood from natural resource-based activities like farming, fishing and gathering from the forest. Every year a high quality of tiger prawn, crab, honey, wax and dry fish export for foreign exchange. Moreover, Sundarban Biosphere Reserve is tried to build up a hub of ecotourism with a noble task of conservation. Wide varieties of mangroves and other flora, amphibians, reptiles, migratory birds, fish and world famous the Royal Bengal Tiger attract the tourists. But lack of local people's participation at large, the objectives of ecotourism as well as its flourishing are confronted with huge challenges. The detail constrains are widely discussed in Chap. 3.

In rural India, till now, land possession is the yardstick for measuring the economic status. The land is very susceptible to natural calamities. Hence, those who possess more cultivable land get more exposure and more likely to migrate (Banerjee et al. 2011). The counter opinion is that those households own either small or large size of land are less likely to migrate compared to those household possess the medium size of the same (Stark and Lucas 1988). No land or small landholders are often unable to bear the migration cost which restrains them to move out. The big

[4]The per capita income in 2013–2014 at current price (Net State Domestic Product) of West Bengal is Rs 70,059. The monthly NSDP is Rs. 5838 (PIB 2015).

Table 6.11 Landholding by households in the Sundarban

	Migrated HH		Non-migrated HH		Total	
Holding size	Frequency	Percent	Frequency	Percent	Frequency	Percent
0.67 hect. (5 bigha) and above	11	5.5	20	10.0	31	7.8
0.27 (2 bigha) – 0.67 hect. (5 bigha)	30	15.0	28	14.0	58	14.5
0.13 (1 bigha) – 0.27 hect. (2 bigha)	26	13.0	42	21.0	68	17.0
0.13 hect. (1 bigha) and below	42	21.0	28	14.0	70	17.5
No land	91	45.5	82	41.0	173	43.3
Total	200	100	200	100	400	100

Source: Field survey data

land owners are rich men. They often do not cultivate the land themselves; lease it out. The middle size class holders have some money to migrate and need more to cultivate or take on lease more land.

The agricultural land in the Sundarban is basically marginal size class (less than one hectare). Around 63% (188,333 hect.) of the total land holding (300,668 hect.) are under this size class (Agricultural Census 2010–11). Moreover, a substantial proportion of households are landless. In the study, 43.3% households reported that they have no cultivable land to their own possession (Table 6.11). Landlessness is higher among the migrated households (46%) than non-migrated households (41%). Nearly 21% migrated households possess less than 1.0 bigha (0.13 hect.) of agricultural land, whereas 14% non-migrated households possess the same. More than 1.0 bigha (0.13 hect.) of agricultural land possession is higher among the non-migrated households. Non-migrated households, therefore, is comparatively well-off and get a certain amount of financial assistance from the land to sustain at the origin. It is one of the reasons for their retention at the origin.

6.4 Social Capital

Social capital refers to the involvement or interaction of people with individuals and institutions, which help to access information regarding jobs and other things. It increases people's trust, aspiration and bonding, ability to work together, wider access to the institutions and participation and ultimately shape the quality and quantity of life (DFID 2001). Social capital in terms of social network or connection, association with formalised groups and aspiration is the foci of the discussion in this section. Though the social connection of migrants with intermediaries and others has been touched upon in the previous chapters, here, it is discussed in details. The aspiration for the urban living of rural people, economic wellbeing and intergenerational aspiration to work in non-farm sectors are also brought in the purview of the discussion.

6.4.1 Social Networks and Connectedness

The social network strongly affects the rational evaluation of the costs and benefits (thesis) of movement (Massey 1990; Todaro 1969; Sjaastad 1962). Having friends, relatives, co-villagers or members from the own community at the destination drastically reduces the migration cost vis-à-vis increase the expected net return that leads to the probability of migration. In the Sundarban, migrants have an excellent individual social network than non-migrants. Almost all of the migrants (99%) had a social connection before they had decided to migrate out. More than one-third (34%) of migrants received information from relatives or friends (Table 6.12). Nearly 30% collected information from co-villagers or neighbours, around 23% from middlemen and nearly 9% contacted the employer himself and vice versa. On the other hand, 73% non-migrants currently do not have any connection outside the area of origin for finding jobs and other helps. Only 21% have relatives or friends outside the Sundarban, and the rest of them almost have no source to find out a job. Their probability of migrating out is restricted by the lack of connectedness.

The social network has significant ($x^2 = 242.08$, $df = 5$, $n = 400$, $p = 0.00$) and strong association (Cramer V = 0.78) with the migration status (migrants or non-migrants) of the respondent. All the sources make a significant difference between migrants and non-migrants to find jobs. No source or asking to find jobs at workplace contributes most ($Adj.R = -14.8$), and it advocates in favour of non-migrants. Middlemen or contractors ($Adj.R = 6.9$) influence most to the migrants moving out of the Sundarban which is followed by co-villagers or neighbours ($Adj.R = 6.7$) and contact himself or contact by employers ($Adj.R = 3.5$).

6.4.2 Cause or Consequence Argument of Social Network

While almost all of the migrants in the Sundarban have a social connection and it sustains over the time, it evokes the argument whether it is a cause or consequence of migration. Once the number of migrants in an area or community reaches a certain level, each migrant expands their social network. After a certain period, all the households of that area or community become closely connected to someone who has migration experience. At that critical level of the social network, migration becomes 'self-perpetuating' (Massey 1990). The social network differently influences the decision making of household belongs to different socioeconomic hierarchy in a community (Findley 1987). When networks are small or limited in size and undeveloped, upper (rich) and lower (poor) class families are more likely to send migrants. With the increase of number of people in the community with migration experience, the middle class becomes predominant. The middle class is the most risk averse. Therefore, they are unlikely to migrate in the absence of secure networks, whereas the poor people have little or nothing to lose, and the upper classes can afford risks (Findley 1987). In the present study, the households with low (Rs. 3000 or below) and moderate monthly income (Rs. 3000–10,000) hold most of

Table 6.12 Sources to find jobs

Sources		Migrants	Non-migrants	Total
Relatives/friends	Count	68.0	42.0	110.0
	Expected count	55.0	55.0	110.0
	% within source	**34.0%**	**21.0%**	**27.5%**
	Adjusted residual	2.9	−2.9	
Co-villagers/neighbours	Count	59.0	9.0	68.0
	Expected count	34.0	34.0	68.0
	% within source	**29.5%**	**4.5%**	**17.0%**
	Adjusted residual	6.7	−6.7	
Middlemen/contractor	Count	45.0	1.0	46.0
	Expected count	23.0	23.0	46.0
	% within source	**22.5%**	**0.5%**	**11.5%**
	Adjusted residual	6.9	−6.9	
Contact himself/contact by employers	Count	17.0	2.0	19.0
	Expected count	9.5	9.5	19.0
	% within source	**8.5%**	**1.0%**	**4.8%**
	Std. residual	3.5	−3.5	
Others	Count	8.0	0.0	8.0
	Expected count	4.0	4.0	8.0
	% within source	**4.0%**	**0.0%**	**2.0%**
	Adjusted residual	2.9	−2.9	
No source/asking to find jobs at workplace	Count	3.0	146.0	149.0
	Expected count	74.5	74.5	149.0
	% within source	**1.5%**	**73.0%**	**37.3%**
	Adjusted residual	−14.8	14.8	
Total	Count	200.0	200.0	400.0
	Expected count	200.0	200.0	400.0
	% within source	**100.0%**	**100.0%**	**100.0%**

Source: Computed from field survey data

the share (92%) of the migrants – 48 and 44% migrants, respectively (Table 6.10). With the increase of migration, the social network is gradually enlarged and becomes stronger. The workers become more secured in new place, and their vulnerability reduces. The middle-class families start to enter into the migration processes to a large number. Whether migration in the Sundarban is considered as a cause or a consequence of social network needs detailed investigation. But it is very clear that social network has a significant influence on out-migration from the Sundarban.

6.4.3 Association with Formalised Group

Members of a formalised group often share mutually agreed views, rules, norms and sanctions. The trustworthy relationship among the members, reciprocity and exchanges endorse co-operation, disburse information and sometimes provide the basis for safety nets of the peasants. It may be the cause of migration or strongly favours retention. In observation of the study, nearly 65% (261 respondents) are not associated with any formalised organisation or association. However, non-migrants are more likely to be attached to the formalised groups compared to the migrants at the origin. These are shared by 41% and 29% for non-migrants and migrants, respectively. Only membership does not portray the real picture of involvement; nature of participation in the institutional activities and taking part in its decision-making are also important one. Though the total persons with attachment (139 respondents or 41%) to the formalised organisation are less, around 58% have 'sometimes' participated in the meeting or procession or other activities. It is followed by around 32% are 'regular or often' participant and 11% have never participated (Table 6.13). Most of the participants (68%) informed that their opinion is valued or respected in the organisations.

A very low level of attachment may be because of people's reluctance due to over engagement for searching a job or maintaining their daily life. Participation also depends on numbers of organisations' availability in their localities. In the Sundarban, there are several NGOs, village clubs, SHGs, cooperative societies, political institutions, religious groups, etc., and they mushroom at the time of any natural calamities. Moreover, the comparatively less number of migrants participating can be attributed to their recurrent migration over the year. Hence, they get very minimal time to get involved in localised actions.

Table 6.13 Participation in organisational activities in the Sundarban

Nature of participation	Migrants		Non-migrants		Total	
	Frequency	Percent	Frequency	Percent	Frequency	Percent
Regular/often	19	33.3	25	30.5	44	31.7
Sometimes	31	54.4	49	59.8	80	57.6
Not at all	7	12.3	8	9.8	15	10.8
Total	57	100.0	82	100.0	139	100.0

Source: Field survey data

6.4.4 Aspiration for Urban Life

Aspiration is one of the essential components in migration dynamics. Aspiration of an individual is influenced by factors ranging from inherent individual characteristics and acquired character traits to the socioeconomic and political environment. People's consciousness about socioeconomic and political opportunities, social network through mass media, the internet, formalised groups or others and the migration experience itself often increase the aspiration. Studies (Czaika and Vothknecht 2012; Carling et al. 2013) suggest migrants have higher individual aspiration than non-migrants even before they choose to migrate. In the study, some selective factors that are most relevant in the context of the Sundarban, such as aspiration for urban life, aspiration to work non-farm sectors and future aspiration for economic wellbeing, are discussed in details.

The Sundarban is located near Kolkata urban agglomeration. Its hinterland is spreading over the districts of South 24 Parganas in which the Sundarban is situated. Even some of the suburbs like administrative blocks of Canning-1, Jaynagar-1 and Mathurapur-1 comprise the Sundarban. Therefore, the islanders aspire for permanently settling either in Kolkata or in the Suburban where urban life lures them with better quality of life and opportunities.

Nearly 75% respondents informed that they would like to live in urban areas if they get any chance in the future (Table 6.14). The ambition is significantly higher ($Adj.R = 4.8$, $p = 0.00$) among the migrants than the non-migrants ($x^2 = 22.42$, $df = 1$, $n = 400$ and $Phi = 0.24$); the shares are 86% and 65%, respectively. A substantial proportion of non-migrants is also willing to migrate to the urban areas that infer the future potential of migration from the Sundarban.

6.4.5 Aspiration to Work in Non-farm Sectors and Future Economic Wellbeing

Lack of modernisation has made farming sectors in India less attractive and an option with low income generating. The young generation of India is too fascinated to work in non-farm sectors with the faster advancement of construction, manufacturing and IT industries. Besides, the comparatively higher social status is associated with the engagement in non-farm sectors rather than their engagement in

Table 6.14 Willingness to settle in urban areas

	Migrants		Non-migrants		Total	
Live in urban areas	Frequency	Percent	Frequency	Percent	Frequency	Percent
Yes	171	85.5	129	64.5	300	75.0
No	29	14.5	71	35.5	100	25.0
Total	200	100.0	200	100.0	400	100.0

Source: Field survey data

the farm sector. The ambition to work in non-farm sectors attracts the rural young generation into the urban centres. In the study, it is observed that respondent's occupation is significantly ($x^2 = 31.17$, $n = 400$, $df = 2$, $p = 0.00$, *Cramer's* $V = 0.29$) different from their parent or guardian's occupation. Around 73% respondents have reported that their parent or guardians are/were engaged in farming and fishing. But, more than half of them (57%) have not preferred to continue with their parent or guardians' occupation, farming and fishing. Non-migrants (55%) are more likely to continue their parent or guardians' occupation (farming and fishing) than migrants (40%). In addition to that, the age of respondents makes a significant influence ($x^2 = 5.71$, $n = 292$, $df = 1$, $p = 0.00$, $phi = -0.14$) in preference for staying parent or guardian's occupation. The young respondents aged 30 years and below have shown (56%) strong unwillingness to continue with their parent's occupation (farming or fishing) than the respondents aged above 30 years (42%); the (young) migrants are more reluctant than (young) nonmigrants. Young people try to explore more income-generating sectors than farming. But older age group has already been settled in their parent's occupation and could not take the risk of leaving the same and exploring the new traits.

Most of the respondents in the Sundarban aspire to improve their present economic condition in the next 5 years. Around 68.0% migrants have informed that they keep hope for a better economic life while improving earning through migration (Table 6.15). This figure is 60% for non-migrants. On an average, nearly 21% respondents are unduly pessimistic about their future economic wellbeing, which is slightly higher among non-migrants (23%).

Ultimately, most of the respondents are optimistic about their future economic improvement. They are highly influenced by urban life and wish to live in urban centres in the future. A generation wise occupational shift is visible. The farmer or fisherman's son does not want to be a farmer or fisherman. The aspiration of the people in the Sundarban leads the huge migration and also indicates the future potential of the same. Moreover, the social network in terms of connectedness to find a job and other information for migration contributes a lot. Therefore, the people of the Sundarban are endowed with social capitals that facilitate the migration from there.

Table 6.15 Aspiration for financial wellbeing in the next 5 years

	Migrants		Non-migrants		Total	
Aspiration	Frequency	Percent	Frequency	Percent	Frequency	Percent
Optimist	135	67.5	119	59.5	254	63.5
Pessimist	36	18.0	46	23.0	82	20.5
Don't know	29	14.5	35	17.5	64	16.0
Total	200	100	200	100	400	100

Source: Field survey data

6.5 Physical Capital

Physical capital encompasses the basic infrastructure, and producer goods needed to support livelihoods (DFID 2001). Infrastructure not only helps to migrate easily from the remote rural to the urban centres, but it also brings the information regarding opportunities where they are thinking to migrate. Thus, physical capital endows the social capital. Moreover, infrastructure facilitates the circular migration result in the greater inflow of remittances at the place of origin. Infrastructure is mainly public goods like affordable transport and communication, adequate drinking water and sanitation facility, clean energy supply, easy access to the health facility and other modern services, such as banking with ATM, internet services and so on. There is some exception like shelter which is most often private property. Energy supply, banking and internet services are accessed for a fee related to use. However, the availability and accessibility of physical capital indicate the standard of living in an area. An area with the poor standard of living not only push the poor people to move out, but the wealthier also try to escape from there.

6.5.1 Transport and Communication

It has already been mentioned that the Sundarban is well connected by road, rail and water network. A total road length is 19,460.75 km, and road density is 4.25 km per sq. km. (B.A.E. & S. 2010) which is far higher than the Indian average, 0.66 km per sq. km. More than half (53%) of the households reported that they enjoy surfaced road facility in front of their house. Around two-third households get ferry or boat facilities at regular basis to commute between the islands. Meanwhile, three local Eastern railway networks, Sealdah to Canning (46 km.), Sealdah to Namkhana (109 km.) and Sealdah to Diamond Harbour (60 km.), reach the remote corners of the Sundarban. The nonavailability of the transport facility round the clock is a great concern of development. Most of the islands are connected to each other by ferry boat (Plate 6.1) or vessel services, and a very few numbers of bridges have been constructed during the last decades. Ferry services are available until 8.30 or 9.00 pm. After that, it is hardly available, arranged on demand basis and imposed with excessive late night charges, Rs. 20–50, whereas the normal fare is Rs. 1.0–2.0 per trip per person. Hence, after 8.30 or 9.00 pm, one island is totally disconnected from others. Every nook and corner of the habitation in the islands gets the mobile facility from almost all the commercial telecom service providers. Nearly 78% households possess at least one mobile connection. The DTH service has gradually been entered into the Sundarban. Around 29% households own television, and 19% households are enjoining DTH facilities. The use of the Internet is negligible. Only 19 respondents, including 15 migrants and 4 non-migrants, access the internet through their mobile and laptop.

Plate 6.1 Ferry Boat, Ferry from one Island to another, Jatirampur Ferry, S 24 Pgs

6.5.2 Energy Usages and Supply

According to the 2011 Census, kerosene is the primary source of lighting in the rural South 24 Parganas. A total of 60% households use kerosene which is far higher than the rural national average (43%). Likewise, around 35% households enjoy electricity, whereas all India average is 55%; about 4% households avail solar cell for lighting in the district of rural South 24 Parganas. From the filed investigation in the Sundarban, it is observed that kerosene, electricity and solar energy usages as a source of lighting by the households are 39%, 36% and 26%, respectively (Table 6.16).

Kerosene is the main source of lighting in the migrated households (41%), and the use of electricity is higher among non-migrated households (42%). The unwillingness to take the connection by the migrated households not only due to the high rate of per unit cost of electricity but also due to mostly the aged, child and females who are left out in the households after the migrating of working people. On the contrary, young people live in non-migrated households, and they aspire for better living and entertainment. However, the availability of electricity for 24 h in rural areas of West Bengal is a great concern. Due to the island set up of the Sundarban, it is very difficult to develop the infrastructure for electrification through the conventional electrical grid. Therefore, most of the households have to depend on either kerosene or solar energy. In addition to that, firewood and crop residue are the main sources of fuel for cooking in the Sundarban. Their shares are around 70% and 21%, respectively. The cow dung cake is used by 7% households. A very few households (2%) who are economically well-off take LPG gas connection. But the consumption is very less, one to two cylinders in a year. People avoid using LPG due to the delivery of the cylinder is very poor quality.

Table 6.16 Source of lighting in the Sundarban

	Migrated HH		Non-migrated HH		Total HH	
Sources	Frequency	Percent	Frequency	Percent	Frequency	Percent
Electricity	62	31.0	83	41.5	145	36.3
Kerosene	81	40.5	70	35.0	151	37.8
Solar energy	57	28.5	47	23.5	104	26.0
Total	200	100.0	200	100.0	400	100.0

Source: Field survey data

Overall, vast area of the Sundarban is still without electricity. Though overall electrification in West Bengal is very low (54.5%) compared to the Indian average (67%) (Census of India 2011), the Sundarban is not only lacking behind the national average but also far away from the state average. A vast area goes under deep darkness just after the sunset. Electricity is highly associated with better quality of living of the households (Das and Mistri 2013). Modern communication technology, such as digital TV, DTH, mobile, laptop, internet service, banking and ATM facilities, etc., are highly dependent on electrification. However, there is a broad scope of clean energy or renewable energy generation, such as solar and tidal energy, in the Sundarban. The utmost willingness of the local administration and the Department of Sundarban Affair (DoSA), Government of West Bengal, is required for that development.

6.5.3 Provision of Safe Drinking Water and Sanitation

Almost all census villages or mouzas (97%) of the Sundarban enjoy the drinking water facility (B.A.E. & S. 2008). The withdrawal of groundwater with hand pump is the primary source of drinking. Nearly 73% of the households are dependent on it. Wherever the fresh groundwater is not available for drinking, tap water from public health engineering (PHE) is the main source (29%). Moreover, most of the households (91%) get drinking water facility within their premises that is 500 m from the house. But some pockets are still suffering from lack of drinking water supply. They travel away from the premises (beyond 500 m) to collect water. Along with quantity of water, quality is also a major concern. The Sundarban is one of the extremely arsenic-affected or red alerted zones (more than 300 μg/L) and iron concentration beyond permissible limits (SOES 2010). Seasonal salinity and declining trend of groundwater level are also associated with it.

Nearly two-third of the households reported that they have their own toilet. Only 7% households share toilet with other family, and 15% households use public toilets. While half of the Indian households and most of the rural households (67%) practice open defecation, the Sundarban has portrayed a distinctive picture, where only 13% households practice open defecation. It is much lesser than the state average that is

39%. At this juncture, the local NGOs' contributions are appreciated in achieving it. The Rangabelia Tagore Society for Rural Development (TSRD)[5] in Gosaba Block, Joygopalpur Gram Vikash Kendra (JGVK)[6] in Basanti Block, Southern Health Improvement Samity (SHIS), Mukti and other local NGOs have been struggling more than a decade for the noble mission. They supply slab and pits, erect toilet, sanitary soap and liquid and make the people aware from time to time.

6.5.4 Availability and Accessibility of Health Facilities

A total of 113 medical institutions, including 2 hospitals, 9 rural hospitals, 8 block primary health centres (BPHCs), 47 primary health centres (PHCs) and 47 N.G.O or private bodies (nursing homes), are dispersed over 5400 sq. km. areas and serve 4.43 million (44.3 lacks) people of the Sundarban (B. A. E. & S., 2013). A total of 158 doctors are available in 2013, and patients per doctors are 10,400 persons. A total of 1,296 beds are available in 113 medical institutions. The medical institutions are not evenly distributed over the islands of the Sundarban. On an average, more than 50,000 people reside in every island. Transport facilities across the islands are poor, not accessible around the clock, especially at night. Even the available mode of transportation, such as engine van, paddle van and boat (Plates 6.1, 6.2 and 6.3), is very slow, and getting an ambulance in an emergency is extremely difficult. Hence, Sundarban people much more depend on the quack doctors at their locality. Around 65% households reported that at the time of illness, family members first meet with quack practitioners with a lot of hope. Only 32% households prefer to visit public medical institutions, and a very few households visit to nursing home or private clinic or registered private practitioners.

The health services are of great agony to every (economic) class of people in the Sundarban. For a simple medical test or a minor health problem, they have to suffer a lot; the availability of quality care lacks everywhere. Not only migrants but those who are government servants or wealthier people also try to leave the Sundarban after retirement and settle down to Kolkata UA for better health facilities. One local primary school teacher from the island of Satjelia in Gosaba Block named Mr. Shyamal Mandal (age 40) has opined that:

> In case of fever, to consult a registered practitioner or for a simple blood test for the same, I have to go to block headquarter, Gosaba which is 14–15 km. away from my village. If someone is suffering a heart attack or a fatal accident, the death is certain. If someone suddenly feels sick at midnight, he has to wait till morning to get quality treatment.

---

[5]A veteran social worker, Padmashri Tusar Kanjilal, is associated with TSRD. He is also awarded with 'Bajaz Peace Prize' for his notable work with the society.

[6]Joygopalpur Gram Vikash Kendra (JGVK) received 'Indira Gandhi Paryavaran Puraskar' in 2010 by MoFE, GoI and 'Biodiversity Award' in 2014 by West Bengal Biodiversity Board, Govt. of WB.

Plate 6.2 Engine van and paddle van, principal mode of road transportation

Plate 6.3 Boat for water transportation

The same tune has been resonated in the voice of a village postmaster in Gosaba Block named Mr. Bishnu Pada Gayen (age 56) that:

> I have three more years to retire. We (me and my wife) are suffering different old age health problems. It is not always possible to go to Kolkata for a medical check-up. Even my son is not able to frequently come to Sundarban and take care of us. My son and daughter-in-law have already shifted to Sonarpur (suburban of Kolkata) where we have made a house. I am waiting for retirement to move there with my wife.

Table 6.17 Types of houses in the Sundarban

	Migrated HH		Non-migrated HH		Total HH	
Type of House	Frequency	Percent	Frequency	Percent	Frequency	Percent
Permanent (pucca)	22	11.0	26	13.0	48	12.0
Semi-permanent (semi-pucca)	85	42.5	89	44.5	174	43.5
Temporary (kutcha)	93	46.5	85	42.5	178	44.5
Total	200	100.0	200	100	400	100.0

Source: Field survey data

6.5.5 Decent Housing

Decent housing is one of the basic three needs, food, cloth and shelter, of human being. The decent housing includes the types of houses[7] (permanent, semi-permanent and temporary), condition of the houses (good, liveable and dilapidated), number of dwelling rooms, having a separate kitchen or not and so on. In the study, more or less equal proportion (44%) of households lives in semi-permanent (semi-pucca) and temporary (kutcha) houses (Table 6.17). Kutcha houses are slightly more prominent among the migrants (47%) compared to the non-migrants (43%). The kutcha houses are mainly made up of mud and paddy straw which are readily available in the Sundarban. On the other hand, only 12% households live in permanent (pucca) houses which are comparatively higher among the non-migrated households (13%). A pucca house in rural area is a sign of economically well-off condition of the households.

Almost half of the households' housing condition[8] is livable, and nearly 22% households' house is in good condition (Table 6.18). But a substantial proportion of households (29%) live in dilapidated houses. It is slightly higher among the non-migrated households (31%) compared to migrated households (26%). Likewise, 26% migrated households live in a house with good condition, whereas nearly 18% non-migrated households live in the same. Overall, most of the houses in the Sundarban are either temporary or semi-permanent and in livable condition; no significant difference is found between the migrated and non-migrated households.

[7]Census of India, 2001 classified the houses into three categories: permanent (*pucca)*, semi-permanent (*semi-pucca)* and temporary (*kutcha*). The present study also followed the same classification and definition. A house with both the wall and roof made of permanent material, such as concrete, asbestos, tiles, burnt brick and tine, is defined as permanent or *pucca* house. Likewise a structure has either the wall or the roof, but not the both, made of permanent material are called as semi-permanent or *semi-pucca* house. A house with both the walls and roof is made of materials that need to be replaced frequently, and material other than those mentioned above is treated as temporary or *kutcha* house.

[8]Those houses which show signs of decay or those breaking down and require major repairs or those houses decayed or ruined and are far from being in conditions that can be restored or repaired are considered as 'dilapidated'. Those houses which require minor repairs are considered as 'livable'. Those houses which do not require any repairs and in good condition are considered as 'good' (Census of India 2001 & 2011).

Table 6.18 Housing condition in the Sundarban

	Migrated HH		Non-migrated HH		Total HH	
Housing condition	Frequency	Percent	Frequency	Percent	Frequency	Percent
Good	52	26.0	35	17.5	87	21.8
Livable	96	48.0	103	51.5	199	49.8
Dilapidated	52	26.0	62	31.0	114	28.5
Total	200	100.0	200	100.0	400	100.0

Source: Field survey data

In addition to that, around 61% households have at least two dwelling rooms,[9] and 31% households have more than two dwelling rooms in their housing unit. The average person per room is 3.5 or around 4.0. Moreover, most of the households that are around 70% use a separate kitchen for cooking which is either in or outside of their houses.

6.5.6 Banking Services

Access to banking facility to every citizen is a part of the inclusionary economic growth. In migration context, it is not only the easiest means of remitting, but it is also a safe or a riskless mode of remitting. Besides, the bank account will also help for 'entitlement portability' of migrants from their place of origin to the destination for the better social security and protection in the near future. Around 62% respondents have recorded to possess a bank account. On an average, one out of every two avails bank facilities, and most of them opened account with zero balance under the central government's flagship programme the Pradhan Mantri Jan Dhan Yojana (PMJDY).

6.5.7 Amenity Index

With the help of the above discussed physical assets, a composite index of amenity is constructed to compare between migrants and non-migrants. The detailed list of the variables and their categories is represented in Table 6.19.

A principal component analysis (PCA) with varimax rotation technique is taken to help construct the index. The three-component solutions explain total 56.2% of the variance, and by taking an average of the component scores, amenity index (AI) is constructed for individual households. All variables are unidirectional. The

[9]A dwelling room includes living room, bedroom, dining room, drawing room, study room, servant's room and other habitable rooms. It does not include kitchen, bathroom, latrine, store room, passageway and veranda which are not normally usable for living. A room, used for multi-purposes such as sleeping, sitting, dining, storing, cooking, etc., is regarded as a dwelling room (Census of India 2011).

Table 6.19 Variables used for construction of the amenity index (AI)

Sl. No	Variable	Category
1	House type	Permanent-1, semipermanent-2, temporary-3
2	Housing condition	Good-1, liveable-2, dilapided-3
3	Person per room ratio	Continuous variable
4	Separate kitchen inside or outside of the house	Yes-1, no-2
5	Source of drinking water within premises (within ½ km)	Yes-1, no-2
6	Source of lighting	Electricity or solar energy -1, kerosene-2
7	Toilet facility	Own toilet-1, share with other family-2, public toilet-3, no toilet/open-4
8	DTH facility at home	Yes-1, no-2
9	Availing banking service by any member of the HH in terms of having a bank account/ ATM card	Yes-1, no-2
10	At the time of illness, where the family members go first to seek treatment	Public health institutions or nursing home/private clinic/registered practitioner- 1, quack practitioner-2, traditional healer-3
11	Enjoy public road facility in front of the house	Metalled-1, unmetalled-2

higher score in subcategories of a variable denotes worsened situation. Therefore, households with higher AI score infer the lower access to different amenities. The formula used for constructing the index is as follows:

$$S_1 = W_1 * Y_1 + W_2 * Y_2 + W_3 * Y_3 \ldots\ldots\ldots\ldots\ldots + W_{11} * Y_{11}$$

where:

S = Score of the component

Y = Standardised variable (individual uptake of facility)

W = Component score coefficient (weightage)

The AI scores vary from a minimum of −1.244 to a maximum of 1.069 (Table 6.20). Almost half of households score 0.00 and below. Around 32% households score in between 0.00 and 0.50 that is the moderate level of uptake of basic amenities. The rest, 18% households, score comparatively good that is 0.50 and above. Independent samples *t*-test examines that there is no significant difference ($p = 0.378$) in the mean score between migrated and non-migrated households in terms of uptake of basic amenities ($t = 0.883$, $df = 398$, $n = 400$, eta squared $= 0.002$).

In general, basic infrastructure in the Sundarban in terms of transport and telecommunication is good. These act as a great incentive for migration from the Sundarban. But, vast areas still lack electrification through the electrical grid that is

Table 6.20 Amenity index (AI)

Statistics	Migrated HH		Non-migrated HH		Total HH	
Mean	0.022		−0.022		0.000	
Median	0.046		−0.045		0.007	
Mode	−0.839		−0.608		−0.839	
Std. deviation	0.493		0.507		0.500	
Minimum	−1.027		−1.244		−1.244	
Maximum	1.069		0.964		1.069	
Skewness	−0.139		−0.034		−0.086	
Observation	200.0		200.0		400.0	
	Migrated HH		Non-migrated HH		Total	
AI Score	Frequency	Percent	Frequency	Percent	Frequency	Percent
≤0.00	93	46.5	106	53.0	199	49.8
0.00–0.50	72	36.0	57	28.5	129	32.3
≥ 0.50	35	17.5	37	18.5	72	18.0
Total	200	100	200	100	400	100

Source: Compute from field survey data

very hard to believe in the contemporary world of information and communication technology (ICT) as well as in the wave of 'digital India'. The lack of electricity is not only the barrier of promoting ICT, but it is also the barrier of expansion of livelihood opportunities at the local level by introducing small or medium industries, household industries, agro-based industries and fish processing industries. The ecotourism is also being hampered due to unavailability of electricity. In hot and humid summer, tourists turn their faces off from the Sundarban. Non-electrification exaggerates the cost of ecotourism. Health facility in the Sundarban is a great agony to all classes of people. A huge proportion still depends on quack practitioners. Though most of the households enjoy the provision of drinking water and sanitation, the quality and quantity of drinking water are a major concern. The housing condition of the islanders infers the poor quality of living. Most of the people live in either temporary or semi-permanent house where the congestion is quite high. On an average, four people live per dwelling unit. Therefore, transport and communication entice the migration; meanwhile, the lack of electricity, health facilities and other poor quality of living provide the extra push for moving out of the Sundarban.

Finally, the chapter can be concluded that social network, aspiration, transportation and communication have a significant influence on migration from the Sundarban. On the other hand, poor quality of living in terms of lack of electricity and cleaner energy usages, health facility, lack of decent housing and other basic amenities generate a huge push to the migration from the Sundarban. In addition to that, households of the Sundarban possess a very minimal amount of assets in terms of land, livestock and durable assets, and very poor services for the livelihood derived from natural resources are the base of the Sundarban. But islanders are endowed with basic human capitals. Most of them are educated, belongs to working age and possess good health. These strongly advocate the people migrating out of the Sundarban in a large proportion for coping with the adverse situation at the origin.

References

Agricultural Census. (2010–11). *Survey data*. Department of Agriculture, Cooperation and Farmers Welfare, Ministry of Agriculture and Farmers Welfare, Government of India (GOI). http://agcensus.dacnet.nic.in/. Accessed 28 June 2018.

B.A.E. & S. (2004, 2006, 2007, 2008, 2011 & 2013). *District statistical handbook, South 24 Parganas and North 24 Parganas*. Department of Statistics and Programme Implementation, The Government of West Bengal. Kolkata: New Secretariat Buildings: Bureau of Applied Economics and Statistics (B.A.E. & S). http://www.wbpspm.gov.in/publications/District%20Statistical%20Handbook. Accessed 22 Nov 2018.

Banerjee, S., Gerlitz, J. Y., & Hoermann, B. (2011). *Labour migration as a response strategy to water hazards in the Hindu Kush-Himalayas*. Kathmandu: ICIMOD.

Camber, R., & Conway, G. R. (1991). *Sustainable rural livelihoods: Practical concepts for 21st century*. IDS Discussion Paper 296. http://opendocs.ids.ac.uk/opendocs/bitstream/handle/123456789/775/Dp296.pdf?sequence=1. Accessed 25 July 2013.

Carling, J., Demba, P., Yassine, M., & Wu, J. (2013). *Migration aspirations in Senegal: Who wants to leave and why does it matter?* European Policy Brief. http://www.eumagine.org/outputs/Policy%20Brief%20Senegal.pdf. Accessed 16 Sept 2015.

Census of India. (1991, 2001 & 2011). *Census tables*. Office of the Registrar General and Census Commissioner, Ministry of Home Affairs, Government of India. CD ROM.

Czaika, M., & Vothknecht, M. (2012). *Migration as cause and consequence of aspiration*. Working Papers-57, International Migration Institute, University of Oxford. http://www.imi.ox.ac.uk/publications/wp-57-12. Accessed 16 Sept 2015.

Das, B., & Mistri, A. (2013). Household quality of living in Indian states: Analysis of 2011 census. *Environment and Urbanization ASIA, 4*(1), 151–171.

DFID. (2001). *Sustainable livelihoods guidance sheets*. London: DFID.

FAO. (2005). *Lifestock sector brief, India*. Livestock Information, Sector Analysis and Policy Branch, Animal Production and Health Division. Rome: FAO. http://www.fao.org/ag/againfo/resources/en/publications/sector_briefs/lsb_IND.pdf. Accessed 3 September 2015.

FAO. (2011). *Guidelines for the preparation of livestock sector review*. Animal Production and Health Guidelines. No. 5. Rome: FAO.

Findley, S. E. (1987). An interactive contextual model of migration in Ilocos Norte, the Philippines. *Demography, 24*(2), 163–190.

IDFC. (2011). *India infrastructure report 2011, water: Policy and performance for sustainable development*. New Delhi: Oxford University Press.

Massey, D. S. (1990). Social structure, household strategies, and the cumulative causation of migration. *Population Index, 56*(1), 3–26.

McLeman, R., & Smith, B. (2006). Migration as an adaptation to climatic change. *Climatic Change, 76*(1), 31–53.

Sjaastad, L. A. (1962). The costs and returns of human migration. *Journal of Political Economy, 70*(5), 80–93.

SOES. (2010). Groundwater Arsenic Contamination in West Bengal – India (20 years study). http://www.soesju.org/arsenic/wb.htm. Accessed 26 Mar 2018.

Spear, A., Jr. (1974). Residential satisfaction as an intervening variability in residential mobility. *Demography, 11*(2), 173–188.

Stark, O. (1984). Rural-to-urban migration in LDCs: A relative deprivation approach. *Economic Development and Cultural Change, 32*(3), 475–486.

Stark, O., & Lucas, R. E. B. (1988). Migration, remittances, and the family. *Economic Development and Cultural Change, 36*(3), 465–481.

Todaro, M. P. (1969). A model of labour migration and urban unemployment in less-developed countries. *The American Economic Review, 59*(1), 138–148.

Wolpert, J. (1966). Migration as an adjustment to environmental stress. *Journal of Social Issues, 22* (4), 92–102.

World Bank. (2013). *India: Diagnostic assessment of select environmental challenges; An analysis of physical and monetary losses of environmental health and natural resources*. Vol. 1, Report No. 70004-IN. Disaster Management and Climate Change Unit, Sustainable Development Department, South Asia Region. Washington, DC: World Bank.

Structures and Processes and Migration 7

Abstract

The structure and processes are essential for the equitable access and distribution of resources, creation of employment opportunities and protection of the entitlement of human beings. Structures and processes are just like the hardware and software to function a system. To develop a backward area like the Sundarban needs a special policy and administrative setup, especially at the governmental level. As the Sundarban is a world biosphere reserve and natural heritage site, it also has a global exposure; hence, various non-governmental organisations at national and international level benevolently work for the development of the Sundarban. In Chapter 7, both public and governmental agencies and their activities are discussed and finally try to link with migration processes.

Keywords

Structure and processes · Department of Sundarban affair · Joint Forest Management Committee · Capital expenditure

Transforming structure and processes are such components which shape the livelihoods, promote equitable access and exchange of capitals and support the multiple livelihood strategies or choices. Structures denote the organisations or institutions which may be private or public. Public organisations include political or legislative bodies at various levels from local (gram panchayat) through to national, executive agencies, such as ministries, departments and commissions, judicial bodies like supreme courts and high courts and other parastatal or quasi-governmental agencies. Likewise, NGOs or private commercial organisations also operate at different levels from the multinational to the very local. Processes refer to the policies, programmes and legislations those are initiated or implemented by the public and private organisations. If structures are hardware, processes are software in the livelihood framework (DFID 2001). Both are inseparable and

A. Mistri, B. Das, *Environmental Change, Livelihood Issues and Migration*,
https://doi.org/10.1007/978-981-13-8735-7_7

very complex nexus between them. Legislation and policies are essential for the equitable access and distribution of resources among the poor, creation of job opportunities, eradication of poverty and deprivation. But without legislative or parliamentary bodies, there is not legislation; without judicial and administrative bodies to implement it, legislation is worthless. Lack of visionary and ground relevancy of policies, development not only goes in vain, but proper structural deficiency is also a major constrain for development, especially in remote rural areas. Most of the cases, organisations' have a tendency to be an urban or a metro centric, hardly stretch their branches or bureaucracy over the rural areas result in services go undelivered or monopolised by the political colours, deprive the needy one, markets do not function properly, and people's overall vulnerability increases and forced to leave out of the area. The Sundarban is an environmentally fragile region, geographically remote located and a habitation of socioeconomically backward people; hence, transforming structures and policies have an important role to the development. The present chapter mainly deals with that. Though some rules and regulations and its associated organisations and activities of some NGOs and public systems have more or less been highlighted during the discussion in previous chapters, here, it is provided insight into some important one with empirical rigour.

7.1 Department of Sundarban Affair

In January 1994, the Government of West Bengal created a new and independent department, Sundarban Affairs, to promote social, economic and cultural advancement of the people residing in the Sundarban areas of the two districts. The Sundarban Development Board (SDB) was placed under this department to implement its development activities in this region. Even, in the British period, a distinct agency, 'Sundarban Commissioner', was designed to look after all the development issues since the early nineteenth century (DoSA 2018). After the independence, the region was monitored by the normal administrative set up (South and North 24 Parganas district administration) until 1973. In 1973, the state government felt the urgency to set up a distinct agency for the development of the backward region and created the SDB under the Development and Planning Department, Government of West Bengal. After the creation of the Department of Sundarban Affair (DoSA), it is placed under the department. SDB functions as a directorate of the department. Its area of functioning pertains to the formulation of integrated programme for effective utilisation of the resources, co-ordination and supervision of execution of plans and review and evaluation of the progress of implementation and making adjustment in policies and measures. Another wing of the DoSA is Sundarban Infrastructure Development Corporation Ltd. (SIDL). It functions as a state government agency under the DoSA. The primary objectives of the DoSA are-

- Expansion of transport and communication system with special emphasis on the construction of high-quality roads, reinforced concrete jetties and bridges and surfaced roads in islands

- Development of irrigation facilities for increasing cropping intensity
- Extension of the power supply through conventional and non-conventional energy sources
- Enhance the livelihood opportunities in agriculture and allied sectors and social wellbeing of the inhabitants

The operational areas of the DoSA are spread over 17 police stations which include 12 police stations from South 24 Parganas and 5 from North 24 Parganas districts. The law and governance of the inhabited part of SBR is under two district magistrates (DMs) in two districts. The RF is divided in two parts – STR which is under a field director and divisional forest (South and North 24 Parganas) which is under a joint director. The fishing and its related issues are governed by the Department of Fisheries, Government of West Bengal. Therefore, there is a serious problem on fund allocation, implementation of schemes and administrative responsibility. Sometimes too many cooks spoil the broth situation observed in the development of the Sundarban. Every work needs the sanctions of two DMs and various BDOs, whose priority are different. The convergence of services of different departments is needed for speedy progress of development processes.

7.2 Joint Forest Management

Under the guideline of joint forest management, Eco-Development Committee (EDC) and Forest Protection Committee (FPC) were formed in 1996. The aim was to involve local people in forest management for effective conservation and create alternative means of livelihoods along with the development of the fringe areas of STR. EDCs and FPCs comprise members from villages extended 1–1½ km towards inland from the fringe. In 2014, EDCs and FPCs have been renamed into Joint Forest Management Committee (JFMCs) issuing a government resolution by the state government. A total of 14 EDCs and 11 FPCs have been registered till date, which cover more than 8584 households in the region. Construction of brick roads, tube well for drinking water, solar cells for lightening, erection of flood centres and community halls, dredging of canals and ponds for irrigation, credit for livestocks, paddle van for engaging in transportation activities, financial assistance to construct pucca houses, purchase pump set and spray machines for agriculture, support self-help groups with EDCs and FPCs zone, job creation through mangrove plantation and conservation, etc. are the major areas of development in the EDCs and FPCs zone. From the field investigation, it is observed that only 12% households in the EDCs and FPCs zone had received any type of assistance from the committees during the last 5 years. A large proportion remains out of a single benefit. Local people reported that the distribution of credits and other endowment is confined to the political ribbon and kinship; the deserving claimants remain underprivileged. Their serious allegation on the system is that most of the development works are carried out on pen and paper and hardly visible on ground. An aged (61 years) fisherman, Mr. Amal Mandal, from village Rajatjubili, Gosaba Block, shares his experience about the EDC activities in his area that:

> If Forest Department renovates the canal, 10–12 contractors are appointed. Contractors are the best beneficiary. Their families live in good condition. They rob most of the amount, and works remain incomplete, or of poor quality. Labourers do not get their wages on time. We are fishermen. We do not get any assistance or it does not reach to the people who need it.

7.3 Cooperative Society, Gramin Bank and Commercial Bank

There is about 100 years' glorious history of the cooperative movement in the Sundarban. It was initiated in 1915 by the father of the Sundarban, Daniel Mackinnon Hamilton. He bought three islands of the Sundarban, namely, Gosaba, Rangabalia and Satjelia, to set up a model estate named 'Hamilton Abad' (Map 7.1). He built up cooperative society, bank, paddy shells, rice mills and stores to sell out

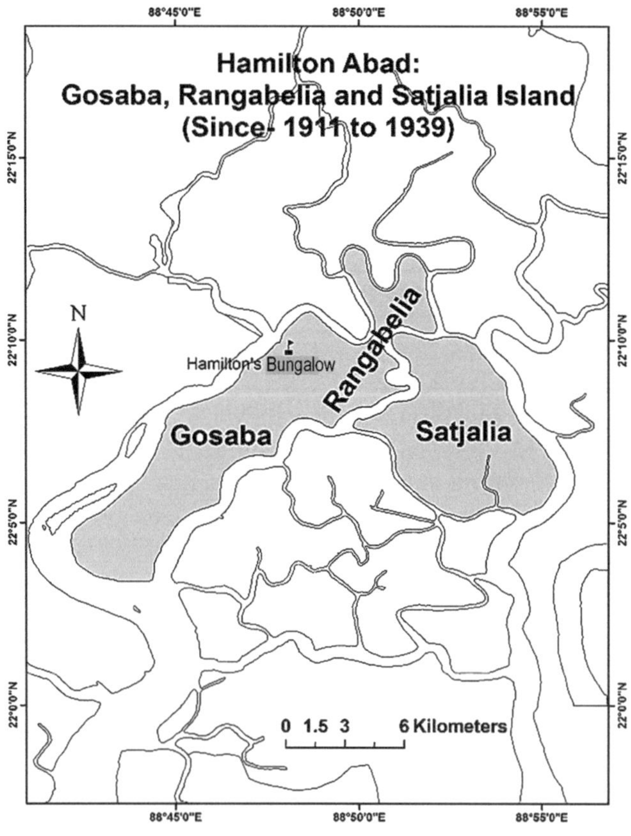

Map 7.1 Hamilton Abad

agricultural products in his estate (Feature Box 7.1). That was the beginning of the cooperative movement in the Sundarban. A total of 138 banks, including 90 Gramin banks and 48 commercial banks, are being operated in the Sundarban. Irrespective of types of banks, one bank serves on an average of 0.64 million (6.41 lakhs) people in the Sundarban (Table 7.1). Banks per lakh persons is around 0.16 which is higher than Indian average 0.01, where developed nations like Germany (2.3), Denmark (2.2), the USA (2.1), France (1.1) and the UK (0.5) (Raghavan 2014) are far higher than all Indian average. However, the matter of concern is the universal accessibility as well as capability to access the banking facility which lacks extremely among the poor people of the Sundarban.

Feature Box 7.1: Daniel Mackinnon Hamilton: Pioneer of the Cooperative Movement

The Sundarban has a glorious history of the cooperative movement. In between 1911 and 1915, a Scottish businessman named Daniel Mackinnon Hamilton bought three islands – Gosaba, Rangabelia and Satjelia – from British Government (Map 7.1). He was a senior partner and chief executive officer of the Mackinnon Mackenzie & PNO Shipping Company in Kolkata (Jalais 2011). His target was to implement the co-operative system in rural India and to show the British Government how much it could be applicable. Hence, he chose these remote islands. That time it was covered with dense forest, marshy low land and two times tidal flow used to inundate the maximum portion of islands, and it was the habitat of the Royal Bengal Tigers, crocodiles and venomous snakes. Every step was full of danger. He deforested the islands; constructed embankment using local saline alluvium soil; managed fresh water for drinking, harvesting by removing saline water and storing rainwater in these places; and set up other basic facilities for habitation. Ultimately, he made the remote islands into habitable one. He invited the farmers to settle here. He leased out these lands on the condition that they would not sell it or lease it out but cultivates it themselves. The farmers came here from remote places such as northern Odisha, the highlands of Bihar and Medinipur and Khulna of Bangladesh (Jalais 2011). He proposed own rule and regulation for his own estate, 'Hamilton abad'. Gosaba was its headquarter.

Till now the name of some places in these Islands carries memories of Hamilton's achievement like Jamespur, Annpur, Luxbagan, Rajat Jubilee and Emilybari. Jamespur was named after Hamilton's nephew, Annpur was named after James's wife and Luxbagan was named after Hamilton's sister Lux. Emily was Hamilton's aunt. Rajat Jubilee was named after King George V's silver jubilee in 1935 (Jalais 2011, p. 41). Besides, some places were also recognised on the basis of its special characteristics like Satjelia. 'Sat' in Bengali is 'seven,' and 'Jelia' are fisherman. Satjelia means 'seven fishermen'. Actually, this place was the habitat of fishermen. Dayapur refers to 'sympathy'

(continued)

Feature Box 7.1 (continued)
or 'kind-heartedness', Sadhupur refers to the residence of 'religious persons' or monk or nun, Charghari means surrounded by 'char' meaning 'sand bar', Arampur refers to 'place of leisure', and Pakhiryala refers to the habitat of birds, especially migratory birds, etc. These names are the local name.

However, he leased out land to the farmer for a certain amount of money. He set up a Cooperative Credit Society with 15 members in Gosaba. He deposited an initial capital of Rs. 500 for the society. In 1918, he started a consumers' cooperative society. In 1919, he established a central model farm for the experiment of paddy, vegetables and fruits cultivation. A cooperative paddy sales society was established in 1923. In 1924, he set up the Gosaba Central Cooperative Bank, and in 1927, he established the Jamini Rice Mill. In 1934, he started the Rural Reconstruction Institute. In 1938, Hamilton printed his own currency which was widely used within his estates for exchange. With his all-out efforts, he made the 'Hamilton Abad' as a well-organised co-operative-driven estate.

Hamilton was an enterprising person. He was well connected with Gandhi and Rabindra Nath Tagore. Tagore came to meet Hamilton on December 1932 in his Gosaba Bungalow (Plate 7.1). He established the 'Gosaba –Bolpure Co-operative Training Institute' whose mission was to train people to lunch co-operative societies in India's rural areas (Jalais 2011). It is heard that central government gave more than 1.0 lakh rupees to this institution in 1938, but this year, Hamilton died. So this mission was not implemented. Hamilton was a kind- hearted man. He left his money and property to the trust created by him. 'Hamilton Trust' is for the sake of the islanders of the Sundarban. After his death, the trust controlled the lands and co-operative societies till independence. But, in 1951, Indian government seized the trust following 'Zamindari abolition laws' on the ground Hamilton's estate was a kind of Zamindari system.

A farmer, namely, Mr. Arup Mistri, age 35 and level of education class-IV, from Gosaba Block stated that:

> I have heard about the getting credit from the Gramin Bank, but never applied for that. It has to face lots of difficulties like fill up the forms, submit the different documents, and has to follow up many times. Even, I have to pay Rs. 2000–3000 in advance to a middleman. Moreover, money is received so late that not to come into actual use.

Likewise, a fisherman, Mr. Provash Sarkar, age 45 and educated up to class X, from Namkhana Block expressed his concern that:

> I am afraid to take loan from Bank. If I will not able to repay on time, the burden of credit becomes excessive. It is better to borrow from the local with comparatively higher interest.

Table 7.1 Cooperative societies, commercial and Gramin Banks in the Sundarban

Co-operative societies				Commercial and Gramin Banks		
Years	Number of cooperative societies	Number of Total members	Working capital ('000 Rs.)	Number of bank offices		Population served/bank offices (No.in' 000)
				Commercial Bank	Gramin Bank	
2000–01	–	–	–	95	45	451
2003–04	698	280,719	453,059	98	45	522
2005–06	699	290,481	587,070	69	46	673
2006–07	700	310,100	839,227	70	46	660
2007–08	681	285,829	577,874	72	45	671
2010–11	691	296,031	643,330	93	48	542
2012–13	717	349,008	1,553,447	90	48	641

Source: B.A.E.& S. (2011)

Plate 7.1 The bungalow in Gosaba where Tagore stayed in 1932

Though the coverage of banking system is much wider in the area, utilisation at the time of need appears to be poor. It is also observed that the formal credit market is competing with the informal market because formal institutions are following cumbersome, slow process. The informal market, though, involves high rate of interest and appears to be more people centric as they reach people on time and more easily.

7.4 Non-governmental Organisations (NGOs)

There are various small and large NGOs actively engaged in social work for the development of the Sundarban. The Rangabelia Tagore Society for rural development (TSRD), Joygopalpur Gram Vikash Kendra (JGVK), Southern Health Improvement Samity (SHIS) and Mukti are the reputed one among them. Provision of safe drinking water, sanitation, women and child health care and education; mobile health facilities in remote areas; supporting handicraft industries; encouraging small savings and investment through SHGs; supply of hybrid seeds, fertiliser and pesticides for farming, sometimes direct cash transfer for livelihood promotion; etc. are the major areas of social work (Plates 7.2, 7.3, 7.4, 7.5, 7.6 and 7.7). Apart from national NGOs, many international agencies extend their hand for the development of the Sundarban. UNESCO is the prime one as the Biosphere Reserve is designated by them. The World Wide Fund for Nature (WWF) gives special attention to the capacity building among the backward people and awareness for wildlife conservation. The UNICEF and WHO carried on different projects on public health. UNDP also funds for the development from time to time. In addition to that, different national and international research institutions and universities have been releasing funds for the research in the Sundarban. Overall, a fair amount of fund is disbursed for the development of the Sundarban from various NGOs to research institutions and universities. During the post disaster period, the fund is increased by leaps and bounds. An in-depth study on NGOs and its activities is essential for the sake of the development in the Sundarban.

Plate 7.2 Flood centre in a Tribal Village, Gosaba, funded by WWF

Plate 7.3 Community climate centre, Gosaba, Funded by SAFE

Plate 7.4 Mobile clinic 'MV Dominique' in the Sundarban

Plate 7.5 Construction of toilet pan by NGO Worker

Plate 7.6 NGO, Mukti Sponsors women SHG, Raidighi

Plate 7.7 A tribal development centre, Raidighi

7.5 Budget and Funding

Capital expenditure incurs on the acquisition of fixed assets (tangible or intangible) which are related to livelihood promotions or enhance the scope of job opportunities and earning profit, improving standard of living and poverty eradication, and the benefit of the expenditure extends to several financial years. The cost of fixed assets in the Sundarban includes brick paved or water bound macadam road and culverts, rainwater harvesting or irrigation and excavation of derelict channel, master sluices and closure, hume pipe sluices, rural water supply, buildings like office-cum-rest sheds, staff quarters and training centres, bridge and jetty constructions, etc. The expenditure whose benefit expires within the year is defined as revenue expenditure. Revenue expenditure incurs for the purpose of floating assets like salaries, wages, rents, rates and taxes; minor works or maintenance; repair and replacement of machinery and plants; conservation and livelihood improvement, such as biodiversity conservation and awareness, social forestry programme, providing fishery inputs and training to the beneficiaries and credit or cash transfer for various projects; etc. A total of Rs. 25.71 billion (2571 crore) has been expensed by the Department of Sundarban Affair, Government of West Bengal, cumulatively for capital (Rs. 12.59 billion) and revenue expenditure (Rs. 13.11 billion) during the last 10 financial years, 2009–2010 to 2018–2019 (Appendix 3). At the very beginning of 2010, capital expenditure for the development in the Sundarban was higher than the revenue expenditure. After that, revenue expenditure has made big stride from Rs. 0.39 million (38.6 core) in 2009–2010 to Rs. 2.39 billion (239 crore) in 2015–2016, whereas capital expenditure has gradually been declined from Rs. 1.38 billion (138 crore) to Rs. 0.47 million (47 crore) during that period (Fig. 7.1). Revenue expenditure was grown into five times higher than capital

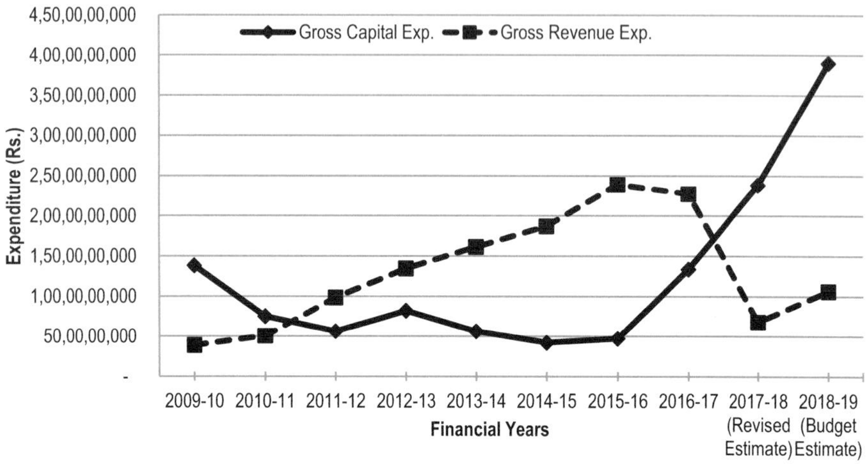

Fig. 7.1 Capital and revenue expenditure in the Sundarban, India (Source: DoSA (2011–2018))

expenditure in the financial year 2015–2016. In 2016–2017, it was changed the strategy to the expenditure for the development in the Sundarban; capital expenditure was tried to rise at 2009–2010 level, and revenue expenditure has begun to decline since then. At present (2018–2019), estimated budget for capital expenditure is Rs. 3.9 billion (390 crore) which is around four times higher than revenue expenditure.

Both the capital and revenue expenditures are essential for the development in an area, but needs to trade-off between them. With the massive revenue expenditure, poverty and underdevelopment can temporally be suppressed, but in long term, it remains in vain. Poor condition of physical capital (infrastructure) as well as quality of life, severe unemployment and low adaptive capacity can be attributed by the low budget allocation for capital formation or expenditure in the Sundarban during financial years. Moreover, effective utilisation of revenue expenditure needs well-administrative setup along with coherent policy frameworks, which are hardly found in the Sundarban, resulting in drain out of huge floating amount and cause of mass grievance among the common people.

Apart from state governmental funding, the central government releases fund for developmental works in the Sundarban through various centrally sponsored or central sector schemes like:

(i) Conservation and management of Sundarban mangroves in West Bengal
(ii) Conservation and management of Sundarban Biosphere Reserve in West Bengal
(iii) CSS Project Tiger from National Tiger Conservation Authority
(iv) Conservation and Management of Sundarban Wetland in West Bengal
(v) A small grant UNDP Project on 'Strengthening Sustainable Livelihood for Biodiversity Conservation in Indian Sundarban'

In addition to that, Mahatma Gandhi National Rural Employment Guarantee Scheme (MGNREGS) is being implemented in the divisions (South and North 24 Parganas) of Sundarban Biosphere Reserve, and the funding for the same is done by District Magistrate's fund.

7.6 Uptake of Credit, Relief Fund and PDS

From the ground investigation, it is observed that uptake of financial support for the livelihood promotion from the governmental and non-governmental institutions or schemes is very minimal. Only 9% households have received credit from the governmental schemes, and around 5% households have received credit from non-governmental organisations during the last 5 years (Table 7.2). But most of the people (87%) have not received any credit either from government or non-governments during the last 5 years. Likewise, only 12% households have received subsidised credit from government projects, programmes and other schemes, and only 2% have received from non-governmental schemes. Here, the non-recipients of subsidised credit are also high that is 86%.

The Sundarban is susceptible to natural calamities because of its geographical location and island set up. On an average, in every 2–3 years, a big natural event, such as cyclone, storm surge, flood or excessive rainfall, are observed in the Sundarban. Latest, on May 2009, a severe cyclonic storm, Aila lashed on and severely ravaged. The central government sanctioned Rs. 478 crore from the National Calamity Contingency Fund (Mukhopadhyay 2009, p. 13). The state government sanctioned Rs. 100,000 for South 24 Parganas and Rs. 500,000 for North 24 Parganas (Chakraborty 2015, p.8). A direct cash transfer was also allocated as Rs. 10,000 per family for the severe cyclone victims and Rs. 5000 for partial victims in both districts. Only 38% households were accounted receiving the allocated mount at the time of survey on December 2014. More than half of the islanders were still waiting for the relief packages after 5 years of the cyclone. Respondents expressed their grievances that there are lots of irregularities and mismanagement to the fund disbursement in post-Aila period. Some of these are stated below:

Table 7.2 Financial assistance received by the households in the Sundarban

	Credits		Subsidised credit		Relief (cash transfer)	
Source	HH	Percent	HH	Percent	HH	Percent
Government	34	8.5	46	11.5	153	38.3
Non-government	18	4.5	9	2.3	2	0.5
Both	–	–	–	–	24	6.0
Non-recipient	348	87.0	345	86.3	221	55.3
Total	400	100.0	400	100.0	400	100.0

Source: Field survey data

- Some families did not get their full allocated amount that is either Rs. 10,000 or Rs. 5000 wherever applicable. Some amount had to be sacrificed for the middlemen or political persons as processing fees.
- There are some cases where one family has received two or more relief packages. After the Aila, Gram panchayat members were responsible for making the list of beneficiaries. They divided a family into two or more households and recorded on the list. For an example, a household with parent and two sons is recorded three households on the list like one household for parent and two households for two sons. Thus, one family was given double or triple relief packages, and some amount from there went to the different pockets of local politicians. As a result, the actual claimants are deprived.
- Relief packages were provided irrespective of economic status. The packages have also been given to those families where family members are already in governmental services.

The public distribution system (PDS) has performed 'enormously well' in Bengal since the introduction of the National Food Security Act in 2013 (Dreze 2016). The Sundarban is not an exception. The feedback from 80% households confirms that the ration is distributed at regular basis over the year. There is no significant difference between migrant and nonmigrant households in terms of uptake of PDS services; these proportions are 83% and 77% households, respectively.

7.7 Job Opportunities through MGNREGA

Mahatma Gandhi National Rural Employment Guarantee Act (MGNAREGA) is a flagship programme of Indian Government (since 2006) for livelihood promotion, especially in rural India. It ensures 100 days guaranteed wage employment in every financial year to every household whose adult members volunteer for unskilled manual work. Like other parts of India, it has also created some job opportunities in the Sundarban. Most of the respondents (92%) in the study possessed the job cards till the date of survey (Table 7.3). The share of migrants and nonmigrants is 90% and 93%, respectively, for the same. In spite of having the job cards, all the job card holders were not attached to MGNREGA work during the last 365 days.

Table 7.3 Work experience of job card holders in MGNREGA

Engaged to MGNREGA	Migrants		Non-migrants		Total	
	Frequency	Percent	Frequency	Percent	Frequency	Percent
Yes	105	58.3	95	51.1	200	55.6
No	75	41.7	91	48.9	166	45.4
Total	180	100.0	186	100.0	366	100.0

Source: Field survey data

They shared awful experiences about the scheme associated with the late payment, irregularity in man-days and the rate of wages. Moreover, getting work in the scheme depends on how close people are or how much they are inclined towards the political power dominating the gram panchayats. MGNAREGA works in the Sundarban are highly emphasised on embankment construction or repairing. The chunks of mud are dug near by the river bed, put into sacks and placed to the embankment. The rate of filling to a sack with mud is Rs. 10–12. If a group of labours fills up 20 sacks, it is recorded 30–40 sacks against the wages of the workers by the supervisor. The extra amount for the extra sacks goes into the pocket of the supervisor. Sometimes, labourers and supervisor mutually share the amount as the amount comes to the account of the labourers. The supervisors are selected from the closed circle of the local panchayat members. According to the MGNAREGA guideline, the male and female workers' ratio is 70:30. The female quota is filled up with the supervisors' family members. When money comes to the labourers' accounts in the local post office, panchayat members contact to the concern post office and persuade the postmaster not to disclose the sanctioned wages or man-days list to the public domain. The rampant corruptions involving the stakeholders are visible in MGNAREGA like all other states. Thus, workers do not fully rely on the scheme; they migrate for searching work out of the Sundarban.

7.8 Livelihood Promotional Index (LPI)

From the above livelihood promotion indicators, a Livelihood Promotional Index (LPI) is constructed with the help of standardisation technique. The particular of the variables is listed in Table 7.4. Higher scores of index refer to the lower uptake of livelihood advancement programmes.

Table 7.4 Variable used for the construction of livelihood promotional index (LPI)

Sl. No	Variable	Category
1	Prior work experience in MGNREGA scheme	Yes-1, No-2
2	HH got PDS facilities at regular basis during the last 365 days	Yes-1, No-2
3	HH received any credit during the last 365 days	Government, NGOs and both-1, No-2
4	HH received any subsidised credit during the last 365 days	Government, NGOs and both-1, No-2
5	HH received (Aila) relief packages	Yes-1, No-2
6	HH received any assistance from EDC/FPC during the last 365 days	Yes-1, No-2

The equation for LPI:

$$\mathbf{LPI = ZV_1+ZV_2+ZV_3+ZV_4+ZV_5+ZV_6}$$

where LPI = Livelihood Promotional Index and ZV = Standardised variable (uptake of programme).

The scores vary in a wide range from a minimum −8.450 to a maximum 5.940 (Table 7.5). The standard deviation (*SD*) is also quite high 3.246. However, there is no significant difference (in mean scores) between migrated and non-migrated households in terms of the LPI ($t = -0.649$, $df = 398$, $p = 0.516$, $n = 400$, eta squared = 0.001).

Overall, the non-governmental institutions like cooperative society, Gramin Bank and commercial bank, and NGOs have an immense contribution to the development of the Sundarban from the beginning to the till date. The government has also designed its administrative setup and initiated new departments, committees and management programmes. DoSA and SDB, EDCs, FPCs and JMCs are its examples. A huge amount of budgetary allocation, financial assistance and relief funds are realised by both the state and central governments in every financial year and at the time of natural calamities. There are lots of irregularities and malpractices observed with the disbursement of governmental funds which have frequently been resonated in the voice of the poor islanders. Flagship programme MGNAREGA is fall flat on face. But PDS is working well which plays a vital role in the food security of the common people. Ultimately, both the migrants and nonmigrants are suffered by the poor quality of services and opportunities generated by the systems. Migrants move out of Sundarban and try to cope up with the situation. But non-migrants try to adopt staying within the system. Their intense attachment with the primary activities

Table 7.5 Livelihood promotional index (LPI)

Statistics	Migrated HH	Non-migrated HH	Total HH
Mean	−0.106	0.106	0.000
Median	−0.569	−0.569	−0.569
Mode	3.470	−2.580	−2.580
Std. Deviation	3.196	3.300	3.246
Minimum	−8.450	−8.450	−8.450
Maximum	5.940	5.940	5.940
Skewness	−0.012	0.017	0.006
Observation	200.0	200.0	400.0

	Migrated HH		Non-migrated HH		Total HH	
LPI Score	N	Percent	N	Percent	N	Percent
−2.00 & below	70	35.0	67	33.5	137	34.3
−2.00–0.00	45	22.5	42	21.0	87	21.8
0.00–2.00	32	16.0	40	20.0	72	18.0
2.00 & above	53	26.5	51	25.5	104	26.0
Total	200	100.0	200	100.0	400	100.0

Source: Compute from field survey data

to explore the maximum benefit, job diversification, maintain the family with lower income, low level of aspiration and network, and maximise the tolerance limit are the most common adaptive strategies in the context of the Sundarban.

References

B. A. E. & S. (2011). *District statistical handbook, South 24 Parganas and North 24 Parganas*. Department of Statistics and Programme Implementation, The Government of West Bengal. Kolkata, New Secretariat Buildings: Bureau of Applied Economics and Statistics (B.A.E. & S). http://www.wbpspm.gov.in/publications/District%20Statistical%20Handbook. Accessed 22 Nov 2018.

Chakraborty, S. (2015). Investigating the impact of severe cyclone Aila and the role of disaster management department – A study of Kultali block of Sundarban. *American Journal of Theoretical and Applied Business, 1*(1), 6–13.

DFID. (2001). *Sustainable livelihoods guidance sheets*. London: DFID.

DoSA. (2011, 2012, 2013, 2014, 2015, 2016, 2017 & 2018). Budget, Department of Sundarban Affair, Government of West Bengal. http://www.sundarbanaffairs.in. Accessed 16 July 2018.

DoSA. (2018). Sundarban development board. Department of Sundarban Affair, Government of West Bengal. http://www.sundarbanaffairs.in/about_us.php . Accessed 6 June 2018.

Dreze, J. (2016, October 18). West Bengal's PDS doing Enormously Well. *The Hindu*. http://m.thehindu.com/news/national/other-states/bengals-pds-doing-enormously-well-says-economist-jean-dreze/article8745603.ece. Accessed 14 Oct 2016.

Jalais, A. (2011). *Forest of tigers: People, politics and environment in the Sundarban*. New Delhi: Routledge Taylor and Francis Groups.

Mukhopadhyay, A. (2009). *Cyclone Aila and the Sundarbans: An enquiry into the disaster and politics of aid and relief*. Kolkata: Mahanirban Calcutta Research Group. https://www.google.co.in/url?sa=t&rct=j&q=&esrc=s&source=web&cd=1&ved=0ahUKEwidpKn0_uTKAhXKkY4KHQxhAP4QFggcMAA&url=http%3A%2F%2Fwww.mcrg.ac.in%2Fpp26.pdf&usg=AFQjCNF9Lpe_3SJnKEOLMtncJhUc93FMsw&sig2=x146LUZV5paS7ho8Ozvgvg&cad=rjt. Accessed 7 Feb 2016.

Raghavan, P. (2014, August 22). India's stunted banking sector: Number of banks per lakh population in Germany is 230 times higher than that of India. *The Times of India*. http://blogs.timesofindia.indiatimes.com/minorityview/indias-stunted-banking-sector-number-of-banks-per-lakh-population-in-germany-is-230-times-higher-than-that-of-india/. Accessed 13 Oct 2016.

Foundation of Sustainable Livelihood and Migration

8

Abstract

A sustainable livelihood copes with and recovers from stressors and shocks, maintains or enhances its capabilities and activities, and provides livelihood opportunities for future generations. Migration is one of the outcomes of a livelihood strategy. A Sustainable Livelihood Approach (SLA) helps one to understand people's livelihoods and their strategies associated with the same. In this chapter, SLA is empirically examined to explore the likelihood of migration from Sundarban under different sets of predictors.

Keywords

Sustainable Livelihood Approach (SLA) · Assets pentagon · Livelihood strategy · Binary logistic regression

The aim of the Sustainable Livelihood Approach (SLA) is to understand people's livelihoods, with which their strategies for migration are associated. SLA links a livelihood strategy with various assets possessed by a family, policies and institutions, and other external threats. It is a very simple approach to study the multidimensional aspects of migration. Since the 1990s, SLA has become a dominant approach in development interventions by major international and governmental agencies, although it first appeared in the research literature in the 1980s. In 1987, "Our Common Future" (the Brundtland Report) by The World Commission on Environment and Development pioneered the concept (Solesbury 2003). At the beginning of the 1990s, the United Nations Conference on Environment and Development (1990) enhanced the concept and advocated a sustainable livelihood as one of the important tools to achieve the goal of poverty eradication. In 1991, Chambers and Conway defined sustainable livelihood and its components, such as capabilities, assets (stores, resources, claims, and access), and activities required for a means of living in their pioneering work on

A. Mistri, B. Das, *Environmental Change, Livelihood Issues and Migration*,
https://doi.org/10.1007/978-981-13-8735-7_8

'Sustainable Rural Livelihoods: Practical Concept for 21st Century'. They also nicely delineated when and how a livelihood becomes a sustainable livelihood, as well as its three dimensions of environmental, social, and economic sustainability. In 1994, Cooperative for Assistance and Relief Everywhere (CARE) introduced Household Livelihood Security as a framework for program analysis, design, monitoring, and evaluation (Krantz 2001).

SLA has been widely recognized internationally since 1997, when it was included in the White Paper of International Development, "Eliminating World Poverty: A Challenge for the 21st Century," by the UK's Department for International Development (DFID 1997). The Department for International Development (DFID) transformed SLA into a policy domain with an overriding aim of poverty eradication. DFID published a series of Sustainable Livelihoods Guidance Sheets in 1999, 2000, and 2001.

Until 2000, most agencies and scholars used SLA as a strategy for poverty alleviation. Later, its scope widened; it was used to examine migration, especially for capturing the multidimensionality of the discourse. The use of SLA for environmental migration is a recent inclusion, but these empirical investigations are very sparse. In 2007, in a study on environmental change and out-migration from Chitwan Valley in Nepal by Massey et al., the empirical exercise of SLA was traced out. In 2008, The International Organization for Migration (IOM) (Kniveton et al. 2008) and the Refugee Studies Centre (Boano and Morris 2008) strongly advocated for the use of SLA in their research series on methodology and policy briefing related to climate change and migration. Likewise, The United Nations Division for Sustainable Development (UNDESA-DSD) (2011) conceptualized SLA to examine the nexus between migration and environmental phenomena from three regions of developing nations: Bangladesh, Tuvalu (an independent island nation in the South Pacific), and the Himalayas.

This chapter is an attempt to examine migration from Sundarban from the viewpoint of climate change with the help of SLA. Here, the DFID framework of SLA is followed. Within the premises of SLA, the influence of each predictor on the strategy of migration from Sundarban is examined, while controlling for the others and finally determining the significant predictors. This model is empirically investigated with the help of a binary logistic regression technique.

8.1 Foundation of the Sustainable Livelihood Approach

SLA is a household-centric model. A family possesses five types of assets—natural, physical, financial, human, and social assets—which comprise the livelihood (Fig. 8.1). These livelihood capitals are not mutually exclusive. Together, they make an asset-pentagon that compensates for any factor that suffers deterioration by stimuli. External influences, such as governmental and non-governmental institutions (structures) and policies (processes), and vulnerability contexts are also considered in the gamut of SLA. Vulnerability contexts include shocks, trends, seasonality, and climatic events and processes. The underlying idea of the model is that when a person's livelihood is threatened by

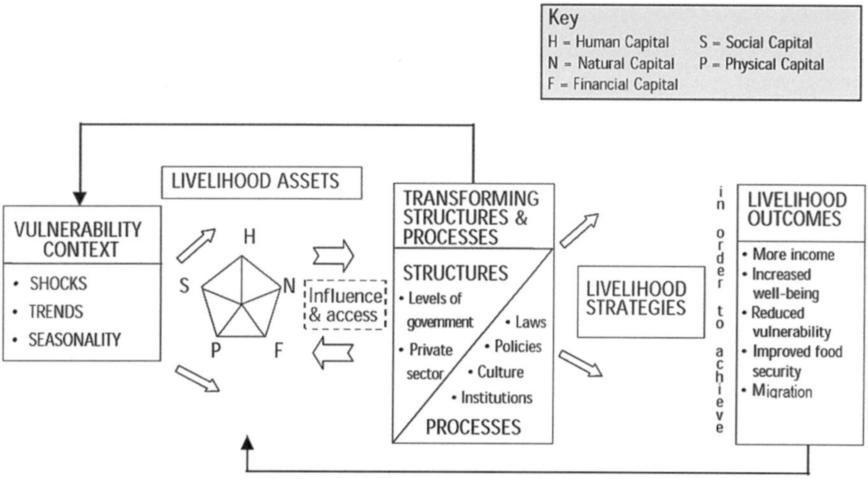

Fig. 8.1 Schematic diagram of the sustainable livelihood approach (SLA) (Source: DFID 2001, p. 13)

stressors, assets help to cope with adverse situations by compensating internally. In this context, the factors related to the structures and processes play a vital role, which influences the efficiency of the assets by enhancing or decreasing the potential utility of the same. Policies and institutions also directly influence the stressors. Based on the interactions between the factors, people can determine their livelihood strategies to cope with an adverse situation. The model also discusses the feedback effect, in which an outcome (strategy) influences the assets or is routed back as an input to become part of the model.

In Sundarban, environmental change is a significant stimulus in the vulnerability context, and out-migration is one of the most prominent livelihood outcomes. The SLA model investigates the probability of migration from Sundarban with reference to non-migrants. Meanwhile, it examines the determinants of migration in Sundarban. Therefore, the model addresses the research question of what factors predict the likelihood of migration in Sundarban. When the environment is considered as a predominant stressor and adversely affects livelihoods, migration from Sundarban turns into an argument of "environmental migration." Hence, the model also helps to examine whether the migration from Sundarban is environmental migration?

8.2 Predictors of Migration

The variable to be examined is the respondents with migration status, which is coded as non-migrants = 0 and migrants = 1. The total respondents or observed cases (n) are 400, which is equally distributed among migrants (exposed group) and non-migrants (controlled group). The probability of migration is investigated for 21 predictors (excluding dummies), which are grouped into seven categories (Table 8.1). The independent variables (IVs) in the model are incorporated from

Table 8.1 List of the variables

Sl. No.	Variables	Description	Category
Dependent Variable (DV): Respondents with Migration Status (non-migrants = 0 and migrants = 1)			
Independent variables (IVs)			
	A. Human capital		
1.	Household (HH) size	Total members of a HH	Discrete variable
2.	Age	Age of the respondent	30 years and younger, older than 30 years
3.	Marital status	Current marital status of the respondent	Married, unmarried
4.	Education	Level of education completed and training /skill of a respondent	Illiterate, up to primary, middle and secondary, above secondary, and other
5.	Work status	Work status of the respondent based on duration of work engagement at the origin	Marginal worker (< 6 months), main worker (≥ 6 months), non-worker
6.	Health condition	Work affected by ill health during last 365 days	Yes, no
	B. Social capital		
7.	Social group	Caste to which a respondent belongs	Scheduled Tribe (ST), Scheduled Caste (SC), non-SC/ST
8.	Social network	Having any connection to finding a job outside of Sundarban	Yes, no
9.	Association with formalized group	Having any membership in a formalized group	Yes, no
10.	Aspiration to an urban lifestyle	Willing to settle in an urban area if provided any opportunity in the future	Yes, no
11.	Generational aspiration to work in non-farm sectors	Preference to continue with parent's occupation	Yes, no
12.	Aspiration for economic well-being	Aspiration for financial success in the next 5 years	Optimistic, pessimistic, don't know
	C. Financial capital		
13.	Household (HH) types	Main source contributing 50% or more of income to the HH	Farming, fishing and NTFPs, daily labor/casual labor, non-farm activities
14.	HH income	Total monthly income (Rs) of a HH	Discrete variable
15.	Livestock of HH	Standard Livestock Unit (SLU) index	Continuous variable
16.	Durable assets of HH	Durable Assets Index (DAI)	Continuous variable
	D. Natural capital		
17.	Agricultural land possession	Amount of land (hectares) possessed by a HH for agriculture on the date of survey	Continuous variable

(continued)

Table 8.1 (continued)

Dependent Variable (DV): Respondents with Migration Status (non-migrants = 0 and migrants = 1)			
	E. Physical capital		
18.	Basic infrastructure and amenities	Amenity Index (AI)	Continuous variable
	F. Structures and processes		
19.	Livelihood promotion by government and non-governmental organizations	Livelihood Promotional Index (LPI)	Continuous variable
	G. Vulnerability context		
20.	Environmental concerns in Sundarban	Risk Perception Index (RPI)	Continuous variable
21.	Risk perception in farming	Risk Perception Index for farming	Continuous variable

different levels, such as the individual, household, and community levels. These are comprised of both metric (discrete and continuous) and non-metric variables (nominal and ordinal). In the case of non-metric variables, dummies are compared with respect to an assigned reference (^{R}) category.

The equation of the model is as follows:

$$\textbf{logit}\,(\boldsymbol{M}) = \ln\frac{\boldsymbol{M}}{(\mathbf{1}-\boldsymbol{M})} = \boldsymbol{\beta}_0+\boldsymbol{\beta}_1*\boldsymbol{X}_1+\boldsymbol{\beta}_2*\boldsymbol{X}_2+\boldsymbol{\beta}_3*\boldsymbol{X}_3+\boldsymbol{\beta}_4*\boldsymbol{X}_4+\ldots+\boldsymbol{\beta}_n*\boldsymbol{X}_n$$

Here,

M = the probability of being a migrant (migrant = 1).
(1-M) = the non-probability of being a migrant (non-migrant = 0)
M/(1-M) = odds ratio of being a migrant
logit (M), or *log (M/1-M)* = *log* of odds ratio
X = IVs/predictors.

8.3 Results and Discussion

Table 8.2 describes the results of the model. The Log likelihood of the model is −84.712 and likelihood ratio (*LR*) chi-square test (Chi^2) value is 385.09. The Log likelihood always becomes negative. Negative and higher values (i.e., closer to zero) of Log likelihood infers a better fit of the model. The logistic regression uses maximum likelihood. The goal of a logistic model is to find the set of parameter estimates that make the data most likely—or in other words, to find values for the

Table 8.2 Sustainable livelihood approach by binary logistic regression

Dependent Variable			*Log likelihood* = − 84.712101		
Respondents with a migration status: non-migrants = 0 and migrants = 1			Observation (N) = 400		
			LR chi^2 (28) = 385.09		
			Prob > chi^2 = 0.0000		
			Pseudo R^2 = 0.6945		
Independent variables	**Category**	**Coefficient (β)**	**Standard Error**	**Odds ratio (*OR*)**	***P***
A. Human capital					
HH size	No. of family members	0.281	0.138	1.324	***0.042***
Age	Older than 30 years[R]				
	30 years or younger	0.791	0.489	2.205	*0.106*
Marital status	Unmarried[R]				
	Married	0.522	0.757	1.685	*0.491*
Education	Illiterate[R]				
	Up to primary	−0.943	0.696	0.910	*0.892*
	Middle and secondary	0.180	0.685	1.197	*0.793*
	Above secondary and other	0.093	0.838	1.097	*0.912*
Work status	Non-worker[R]				
	Marginal worker	−0.778	0.820	0.459	*0.343*
	Main worker	−2.100	0.826	0.122	***0.011***
Health condition	No[R]				
	Yes	0.232	0.495	1.261	*0.640*
B. Social capital					
Social group	Non-SC/ST[R]				
	ST	1.775	0.823	5.903	***0.031***
	SC	0.682	0.553	1.978	*0.217*
Social network	No[R]				
	Yes	3.640	0.065	38.075	***0.000***
Associated with formalized group	No[R]				
	Yes	−1.439	0.492	0.237	***0.003***
Aspiration to an urban lifestyle	No[R]				
	Yes	0.588	0.559	1.801	*0.292*
Generational aspiration to work in non-farm sectors	No[R]				
	Yes	0.751	0.471	2.118	*0.111*
Aspiration for economic well-being	Don't know[R]				
	Optimistic	0.657	0.710	1.929	*0.355*
	Pessimistic	0.411	0.825	1.509	0.618

(continued)

Table 8.2 (continued)

Dependent Variable			*Log likelihood* = – 84.712101		
C. Economic capital					
HH types	Non-farm activities[R]				
	Farming	−1.365	0.595	0.255	***0.022***
	Fishing (including NTFPs)	−1.469	0.680	0.230	*0.031*
	Daily labor/ causal labor	2.928	0.907	18.683	***0.001***
HH income	Rs./month	−0.0001	0.000	1.000	*0.157*
Livestock of HH	Standard Livestock Unit (SLU) Index	0.125	0.316	1.133	*0.692*
Durable assets of HH	Durable Assets Index (DAI)	0.199	0.485	1.221	*0.681*
D. Natural capital					
Agricultural land (ha.) possession		0.375	0.862	1.454	*0.664*
E. Physical capital					
Basic infrastructure and amenities	Amenity Index (AI)	−0.339	0.580	0.713	*0.559*
F. Structures and processes					
Livelihood promotion	Livelihood Promotional Index (LPI)	−0.216	0.078	0.805	***0.005***
G. Vulnerability context					
Environmental concerns in Sundarban	Risk Perception Index (RPI)	0.949	0.287	2.582	***0.001***
Risk perception in farming	Risk perception index for farming	−1.186	0.427	0.306	***0.005***
Constant		−6.103	2.200	0.002	*0.006*

HH-Household, [R]-Reference category, $p < 0.05$- 5% & $p < 0.01$-1% level of significance
Source: Computed from Field Survey Data

parameters (coefficient) that maximize the value of the likelihood function. The "maximum value of the likelihood function" refers to the highest value that is closer to zero. The number in the parenthesis in *LR* $chi^2(28)$ indicates the degree of freedom (*df*). Here, the (*df*) is 28. The *p*-value of LR Chi^2 is 0.000, which is <0.05 and rejects the null hypothesis (H_0) at a 95% confidence interval (*CI*). The *p*-value (0.000) infers that overall the model is statistically significant. The *Pseudo* R^2 indicates that nearly 70% variance within the dependent variable is explained by the predictors.

The significant predictors ($p < 0.05$) in the model are the number of family members, work status of the respondents, caste, social network, association with formalized groups, household types, livelihood promotion by governmental and non-governmental agencies, environmental concern, and risk perception in farming. The household size, for which was recorded $OR = 1.324$ and $\beta = 0.281$, infers that the odds of migration increase by a factor of 1.324 (or 1.324–1 = 0.324, around 32%) for every additional member in a household (Feature Box 8.1). Household size creates pressure on earning members of a family, which lead them to make the decision to migrate. Studies (Massey 1990; Stark and Lucas 1988; Speare 1974) suggest that large family size diversifies the labor portfolio through migration and helps families to cope with adverse situations.

The work status of the respondent (dummy: main worker) based on the duration of work engagement at the origin makes a significant contribution to the model. A respondent who is a main worker (works more than 6 months at the origin) has migration odds that are approximately 88% (0.122–1 = −0.878) less than that of non-workers.

The social network is the strongest predictor in the model. For every one-unit increase of social network—that is, having an additional social connection to find a job (compared with not having the same), an increase of migration by 3.640 log odds is expected while other predictors are constant. The social group in terms of the caste of the respondent is also a significant predictor under social capital. The likelihood of migrating among the Scheduled Tribes (ST) is nearly six times higher ($OR = 5.903$) than that of the non-SC/ST(^{R}) and approximately three times higher than that of Scheduled Caste (SC) ($OR = 1.978$). The higher probability of moving out of an ST community raises the question of social inequality in terms of asset possession and access to other provisions. An association with a formalized group at the origin ($\beta = -1.439$) favors non-migration, leading to less likelihood of respondents reporting as migrants. Migrants are less likely to get involved with any formalized group due to their usual absence from their origin.

Feature Box 8.1: From Probability to Odds to Log of Odds

The odds ratio ($p/(1\text{-}p)$) is a ratio of the probability of success (p) of an event over the probability of the failure ($1\text{-}p$). Probability ranges from 0 to 1, and an odds ratio (*OR*) ranges from 0 to positive infinity. Odds increase as the probability increases or vice versa (Table 8.3).

$$\textbf{Odds Ratio} = \frac{p}{1-p}$$

The Log (ln) of Odds is the Coefficient (β).

(continued)

Feature Box 8.1 (continued)

Table 8.3 Transformation from probability to odds to log of odds

p	odds	log odds
.001	.001001	−6.906755
.01	.010101	−4.59512
.15	.1764706	−1.734601
.2	.25	−1.386294
.25	.3333333	−1.098612
.3	.4285714	−.8472978
.35	.5384616	−.6190392
.4	.6666667	−.4054651
.45	.8181818	−.2006707
.5	1	0
.55	1.222222	.2006707
.6	1.5	.4054651
.65	1.857143	.6190392
.7	2.333333	.8472978
.75	3	1.098612
.8	4	1.386294
.85	5.666667	1.734601
.9	9	2.197225
.999	999	6.906755
.9999	9999	9.21024

$$\beta = \ln\left[\frac{\boldsymbol{p}}{\boldsymbol{1-p}}\right]$$

Therefore, Exponential (exp.) of the β is Odds. **Odds = exp. (β).**

The Coefficient (β) or Log of Odds infers the relationship between an independent variable and a dependent variable. More precisely, it infers the expected change in the log of odds (β) of the dependent variable for every one-unit change of the independent variable, holding all other predictors constant. Greater odds result in greater log of odds and vice versa (Fig. 8.2). If *OR* is >1.0, then the probability is >50%, the β value would be positive nonzero, and vice versa.

(continued)

Feature Box 8.1 (continued)

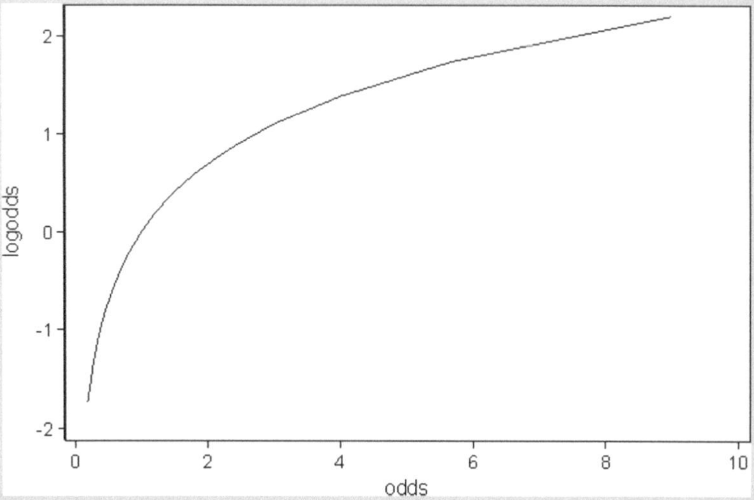

Fig. 8.2 Relationship between log odds and odds

Apart from the social network, social group, and association with a formalized group, an aspiration to an urban lifstylee ($\beta = 0.588$), generational aspiration to work in the non-farm sectors ($\beta = 0.751$), and future aspiration for economic well-being ($\beta_1 = 0.657$ and $\beta_2 = 0.411$) under the social capital argue in support of out-migration from Sundarban; however, the coefficients of these predictors are not statistically significantly different from zero at a 95% *CI*. Household type is the second strongest predictor in the model. The household type is characterized according to the source of income that contributes ≥50% income. The variable is comprised of three dummies—farming, fishing (including NTFPs), and daily or casual labor, where the reference category (^{R}) is non-farm activities. Daily or casual labor ($\beta = 2.928$) exerts a huge push toward migration, whereas farming ($\beta = -1.365$) and fishing ($\beta = -1.469$) deter it. Thus, if the people of Sundarban receive enough support from their principal activities of farming and fishing to maintain their livelihood, then the chances of migrating will decrease.

Daily or casual labor influences the out-migration in two ways. Most of the migrants are attached to daily or casual labor either at the origin or destination. When they are attached at the origin, they suffer extreme marginality (getting a job for less than 6 months) that pushes them to migrate out. When they are attached to the destination, they earn more compared to the origin that pulls the migration. The *OR* of the Livelihood Promotional Index (LPI) is 0.805 and β is -0.216. For each one-unit increase in the LPI score, the odds of migrating decrease by a factor of 0.805. Therefore, if the scope of maintaining livelihood increases, the probability of migration will decrease by approximately 20%.

The environment is the major threat to people's livelihoods in Sundarban. People's overall concern regarding the changing environment in Sundarban positively affects the likelihood of migration. For a one-unit increases in the risk perception index (RPI) score, migration increases by three times ($OR = 2.582$). However, perceived risk in the activities for means of living, especially farming, has shown a negative relationship with the probability of migration ($OR = 0.306$ or $\beta = -1.186$). In Sundarban, a substantial proportion of farmers are landless, and farmland is marginal in size holding. Most non-migrants reported greater concern about farming than did the migrants. Hence, the mean score ($\bar{x}$) of perceived risk in farming for non-migrants ($\bar{x} = 3.01$, $SEM = 0.035$) is higher than that of the migrants ($\bar{x} = 2.88$, $SEM = 0.039$). This phenomenon results in an odds ratio that favors non-migration. Another principal activity in Sundarban is fishing; however, fishing was excluded due to the low number of observations ($n = 268$). However, the influence of the environment is certainly significant, although ambiguous.

Finally, it can be concluded that social capital, such as social network, social group (caste), and association with a formalized group, contribute the most to the SLA model. The issues related to occupation at the origin, such as sources of income (household types), duration of work availability (work status), and livelihood promotion by institutions, play crucial roles in the likelihood of migration from Sundarban. Age, marital status, and education do not contribute significantly, but household size contributes a lot to migration. Socio-economic conditions and housing conditions did not differ much between migrant and non-migrant households, despite receiving a quite substantial amount of remittances in the households. It is possible that, in the absence of migration, a family may be pushed into abject poverty.

Lastly, the influence of environment cannot be denied; it is significant, but ambiguous. Two variables interact in two different ways. In Sundarban, coping with environmental stressors is very slow and age-old practices with the slow changes in the environment. The environmental stimuli in the model are not predominant or convincing reasons to compel people to move out of Sundarban. The question of 'environmental migration' is somehow downplayed. The environmental stressors, therefore, can be considered as additional influences along with the socioeconomic, political, and demographic factors that play major roles.

References

Boano, C., & Morris, T. (2008). Environmentally displaced people understanding the linkages between environmental change, livelihoods and forced migration. In *Forced migration policy briefing 1*. Oxford: Oxford Department of International Development.

Camber, R., & Conway, G. R. (1991). *Sustainable rural livelihoods: practical concepts for 21st century*. IDS discussion paper 296. http://opendocs.ids.ac.uk/opendocs/bitstream/handle/123456789/775/Dp296.pdf?sequence=1. Accessed 25 July 2013.

DFID. (1997). *Eliminating world poverty: A challenge for the 21st century* (White paper on international development). London: DFID.

DFID (1999, 2000 & 2001). *Sustainable livelihoods guidance sheets*. London: DFID.

Kniveton, D., Schmidt-Verkerk, K., Smith, C., & Black, R. (2008). *Climate change and migration: Improving methodologies to estimate flows*. Geneva: IOM.

Krantz, L. (2001). *The sustainable livelihood approach to poverty reduction: A introduction*. Division for Policy and Socio-Economic Analysis, Swedish International Development Cooperation Agency. Stockholm: SIDA.

Massey, D. S. (1990). Social structure, household strategies, and the cumulative causation of migration. *Population Index, 56*(1), 3–26.

Massey, D., Axinn, W., & Ghimire, D. (2007). *Environmental change and out-migration: Evidence from Nepal*. Population studies center research report 07-615. http://www.psc.isr.umich.edu/pubs/pdf/rr07-615.pdf. Accessed 18 Nov 2016.

Solesbury, W. (2003). Sustainable livelihoods: A case study of the evolution of DFID policy. Overseas Development Institute, working paper 217, London: Overseas Development Institute.

Spear, A., Jr. (1974). Residential satisfaction as an intervening variability in residential mobility. *Demography, 11*(2), 173–188.

Stark, O., & Lucas, R. E. B. (1988). Migration, remittances, and the family. *Economic Development and Cultural Change, 36*(3), 465–481.

UNDESA-DSD (2011). *Background paper: Environmentally induced migration and sustainable development*. The United Nations Division for Sustainable Development (UNDESA-DSD). http://www.webmeets.com/files/papers/eaere/2011/1107/Environmentally%20Induced%20Migration%20and%20Sustainable%20Development_Milan_Areikat_Afifi.pdf. Accessed 29 Oct 2016.

9 Environmental Migration and Development Deficit: A Debate

Abstract

Environmental migration suffers not only methodological abstruseness and data limitations but also definitional ambiguity. It is a very policy concern that attribution of environmental migrants to a group without assessing the individual effect size of the stressors of migration and nexus among them plays down the seriousness of real issues vis-à-vis evoking the politics – national and international. Nowadays, it is a very conscious political trick to relate the natural calamities with climate change and hush up the development deficits. Climate change is a matter of concern or hype to hide the shortfall of adequate adaptive measures as well as achieving the unethical political intentions; it is needed to be justified very consciously.

Keywords

Climatic change and variation · Development deficit · Vulnerability · Adaptive measures

Migration from the Sundarban is a predominant livelihood strategy. Whether it is environmental migration or not has been addressed with rigorous data work and statistical exercises in this book. Until now, assessing environmental migration suffers not only methodological abstruseness and data limitations but also definitional ambiguity. It is consensus that when people are obligated to leave their habitual homes 'for compelling reasons of sudden or progressive change in the environment that adversely affects their lives or living conditions' (IOM 2009) is defined as environmental migration. 'Lives or living conditions' that is 'livelihood' is affected by different stimuli including not only environmental stressors but also social, economic, political and demographic aspects and institutions and policies with various degree. In what degree or strength of environmental change compared to the (strength of) other influencing factors can be attributed as environmental

A. Mistri, B. Das, *Environmental Change, Livelihood Issues and Migration*,
https://doi.org/10.1007/978-981-13-8735-7_9

migration is not clear. If a miniscule change of environmental parameters along with socioeconomic and political factors which exert the most for migrating out, or a change of the environment owing to the long-term turmoil of socioeconomic and politics that lead to exploitation of the environment can it be defined as environmental migration? Even 'compelling reasons' of change in the environment are very ambiguous. In social sciences, to assess the degree of effect (effect size) of individual stressor on migration is very complicated and challenging also. Meanwhile, it is a very policy concern that attributing a mobile group as environmental migrants without assessing the individual effect size of the stressors of migration and nexus among them downplay the seriousness of real issues vis-à-vis evoking the national as well as international politics. Nowadays, it is a very conscious political trick to relate the natural calamities with the climate change and hush up the development deficits. Any vulnerability has two facets – exposure and adaptive measures (McLeman and Smith 2006). India is highly exposed to the climatic hazard on multiple fronts. The Himalayas with its vast glaciers fortifies in the North and Northeast; around 70 million hectares of forest areas are endowed with rich mineral resources and habitat of indigenous people; more than 50% people still depend on primary activities; it has around 7000-km-long coastline with 14.2% of total population and 13 metro cities which are the hubs of Indian economy. In addition to that the illiteracy, poverty, caste and creeds, religious and superstitious are embedded in the socioeconomic life of India. Hence, India has to be strengthened in adaptive capacity utilising its vast resources, but till now, it is far from satisfactory. It is critically justified that on an average, more than 800 tornados lash on the USA and cause 80 deaths and 1500 casualties in a year (NOAA 2012). Whereas, in India, 2–3 time less intensity tropical cyclones, for examples Super Cyclone in Odisha with velocity 260 km./hr. in 1999, Sever Cyclonic Storm, Aila with speed 120 km./hr. in West Bengal in 2009, compared to the intensity of the tornados in USA caused deaths and casualties of thousands of people and billions of economic losses (Mistri 2012). Likewise, every year an average 1500–2000 earthquakes with Richter scale 6.5 occur in Japan, and a very few people and property are lost. In 26 January 2001, an earthquake with Richter scale 7.6 occurred in Bhuj in Gujarat. Nearly 19,727 people were dead, 166,000 people were injured, 600,000 were homeless and 348,000 houses were ruined. Total economic losses were 1.3 billion dollars (CIRES 2001). In the same year, on 13 January, an earthquake with the same intensity occurred in Salvador in the USA (GCP 2003). Only 600 people were dead, 2400 were injured and 4500 houses were collapsed. Not only tropical cyclone and earthquake but also floods related to heavy rainfall, such as very recent occurrence (in 2018) of flood in Kerala, flash flood in Leh in 2010, flood in Mumbai in 2005 and many more, portrayed the naked face of the adaptive measures in India. This phenomenon, somehow, subdues the debate of climate change. Climate change is a matter of concern or a hype to hide the shortfall of adequate adaptive measures as well as achieving the unethical political intentions; it is needed to be justified very consciously. Debate of climate change is also pertaining to the data limitation and rate of change of the environmental parameters. During the twentieth century, air temperature has increased 0.74 °C, and mean sea level (MSL) has risen 0.15 metre (IPCC 2007). It is estimated that, during the twenty-first century, MSL will rise 0.18–0.59 metre (IPCC 2007). But it highly depends on how much water will be released through the melting of glaciers that is totally uncertain (Bindoff et al. 2007). In the Sundarban, whatever data set is used in support climate change, most of the cases, not more than 30 years or a climatic normal. Even some parameters have also been registered positive changes. Total forest covers have been increased and densified; wild animals have been increasing. It evokes the dualism between 'climatic variation' and 'climate change' in the Sundarban. The present study does not concentrate on the debate. At the very first, it is assumed that there has been a degradation or deterioration of environment that may be short-term climatic variation or long-term climate change. The study only focuses on subjective appraisal of this change and its effect on livelihood, result in migration as an outcome. However, definitional ambiguity and methodological complexity make the discourse of environmental migration indistinctness. The present study is an endeavour to dissolve the haziness of the discourse in the context of Sundarban Biosphere Reserve in India. The influence of environment on the livelihood of the Sundarban is not denied. The sudden or progressive changes in the environment adversely influence the activities for means of living in the Sundarban, and people are also concerned about it. But its degree of influence is not as strong as socioeconomic and demographic factors compelling the people migrating out of the Sundarban. Thus, the reason behind environmental migration is subordinated here. It can be considered as an additional push along with the socioeconomic and demographic factors which play the major role.

Forest conservation and livelihood conflict are not new in Indian context, where climate change acts as a new instigator. In the name of climate change and conservation, the authorities have gradually been encroaching to the rights of fisherfolks in the Sundarban. They indirectly discourage the decades-long traditional practice of fishing and gathering. Gradually implementing new rules and regulations, stringent the existing one, diminishing the area under fishing (buffer areas), delaying men in renewal of licences or permits, harassing the fishers' fishing ground and so on are the reflection of attitude of the authorities. The Sundarban is not only the territory of the Royal Bengal Tiger but also a habitat of 4.4 million people that needed to understand and be utmost careful to their livelihood. Highly skewed development towards environmental conservation without balancing with economic and social counterparts, the conservation processes will remain in vain. The establishing of Biosphere Reserve by UNESCO (in 2001) is one of the prime causes of that. It provokes a sustainable development where local people participate to the conservation processes and sustainable use of the resources. But the authority's intense conservation interest bypass through rigorous exercises of regulations and keep away local people from the forest in the name of conservation. As a result, a group of fishermen have been compelled to divert from fishing.

Finally, the study evokes some recommendations for the development of the Sundarban. The Sundarban is included in two districts of West Bengal, and too many authorities are entrusted for the development of the Sundarban. There is hardly coordination among them. For the better development, good administration and research, a total of 9600 sq. km. Sundarban Biosphere Reserve should be designed as a distinct administrative unit (district). Migration from the Sundarban is a win-win solution – it helps to the economic development in one hand and forest conservation by reducing the pressure on natural resources on the other hand. A coherent policy framework to facilitate the migration from the Sundarban, ensure the social protection at the destinations and create an investment environment of the remittances is an utmost need in this context. Meanwhile, infrastructure development, especially electrification with the electrical grid in the entire Sundarban and connectedness of the islands with bridges or vessel, is needed to be emphasised. Electrification will help to progress household industries, agro- and forest-based industries, fisheries and fish processing industries and so forth. Interisland interconnectedness will make the transport system fast, frequent and available for round the clock. Improving transportation and electrification will open the door for modernisation and economic inclusion in the Sundarban. Ecotourism has an enormous potential in the Sundarban. Ensuring the participation of local people in large brings them in the forefront of decision-making, and retaining the revenue among them will have to be prioritised in the policy formation. Improving the farming and fishing ensures the sustainable supply of food or the food security to the island areas like the Sundarban, which is the major challenge in the context of global warming. If most of the remittances are drained out for meeting the basic needs (food and shelter), very minimal will remain to invest for further development. The expansion of irrigation and excavation of derelict channel, upgrading the sluices and closures, harvesting the salt-tolerant seeds and facilitating the credits at low interest are the major recommendations for farming. In fishing, the authorities should be more accountable to the entitlement of the fisherfolks and promote a balance development. Moreover, there is a wide allegation against the distribution of governmental funds and livelihood opportunities among the locals. It needs utmost care of policies intervention and restructured or reform the institutions. The Sundarban is highly exposed to natural hazards due to its location and morphological set up. So, adaptive capacity is the best option to cope with the adverse situation. A total of 3500-km-long embankment is the lifeline of the Sundarban. The first and foremost adaptive capacity building should be begun with strengthening the embankment with suitable technology input. Advancement of the warning system and proper knowledge building to cope with different hazards are the main proactive measures. Meanwhile, some reactive measures like quick evacuation, repairing the embankment, rehabilitation, relief and compensation required an immediate basis. Lastly, to implement the above-mentioned recommendations, there is a need of not only a huge funding, but also proper functioning of the systems is essential.

References

Bindoff, N., Willebrand, J., Artale, V., Cazenave, A., Gregory, J., Gulev, S., Hanawa, K., LeQuéré, & Co-authors. (2007). Observations: Oceanic climate change and sea level. In S. Solomon, D. Qin, M. Manning, Z. Chen, M. Marquis, K. B. Averyt, M. Tignor, & H. L. Miller (Eds.), *Climate change 2007: The physical science basis* (pp. 385–432). Cambridge: Cambridge University Press.

CIRES. (2001). 26 January 2001 Bhuj Earthquake, Gujarat, India. http://cires.colordo.edu/~bilham/Gujarat2001.html. Accessed 26 July 2011.

GCP. (2003). Central American Earthquake. http://noosphere.princeton.edu/quake010113.html. Accessed 26 July 2011.

IOM. (2009). *Migration, environment and climate change: Assessing the evidence*. In F. Laczko, C. Aghazarm (Eds). Geneva: IOM.

IPCC. (2007). *Climate change 2007: Climate change impacts, adaptation and vulnerability*. Delhi: Cambridge University Press.

McLeman, R., & Smith, B. (2006). Migration as an adaptation to climatic change. *Climatic Change, 76*(1), 31–53.

Mistri, A. (2012). জলবায়ু পরির্বতন, আভযিে াজন এবং উন্নয়ন (Climatic change, adaptation capacity and development). *Antorjatik Pathsala, 2*(3), 257–264.

NOAA. (2012). *Tornados nature's most violent storms*. http://www.nssl.noaa.gov/edu/safety/tornadoguide.html. Accessed 15 Jan 2012.

Appendices

Appendix 1

To be included on the UNESCO's World Heritage List, a site must be of outstanding universal value and meet at least one out of ten selection criteria. The selection criterions made by UNESCO are as follows:

(i) To represent a masterpiece of human creative genius.
(ii) To exhibit an important interchange of human values, over a span of time or within a cultural area of the world, on developments in architecture or technology, monumental arts, town planning or landscape design.
(iii) To bear a unique or at least exceptional testimony to a cultural tradition or to a civilization which is living or which has disappeared
(iv) To be an outstanding example of a type of building, architectural or technological ensemble or landscape which illustrates (a) significant stage(s) in human history.
(v) To be an outstanding example of a traditional human settlement, land use or sea use which is representative of a culture (or cultures) or human interaction with the environment especially when it has become vulnerable under the impact of irreversible change.
(vi) To be directly or tangibly associated with events or living traditions, with ideas or with beliefs and with artistic and literary works of outstanding universal significance. (The committee considers that this criterion should preferably be used in conjunction with other criteria)
(vii) To contain superlative natural phenomena or areas of exceptional natural beauty and aesthetic importance.
(viii) To be outstanding examples representing major stages of earth's history, including the record of life, significant ongoing geological processes in the development of landforms or significant geomorphic or physiographic features.
(ix) To be outstanding examples representing significant ongoing ecological and biological processes in the evolution and development of terrestrial, fresh water, coastal and marine ecosystems and communities of plants and animals.

A. Mistri, B. Das, *Environmental Change, Livelihood Issues and Migration*,
https://doi.org/10.1007/978-981-13-8735-7

(x) To contain the most important and significant natural habitats for in situ conservation of biological diversity, including those containing threatened species of outstanding universal value from the point of view of science or conservation.

Appendix 2

The details of the reasons for the migration are stated below. These definitions are followed by the National Sample Survey (NSS), 2007–2008 (2012), 64 round.

In search of employment: The persons who were not already in employment at the time of leaving the place of origin and migrated to another place in search of employment are considered as migrated in search of employment.

In search of better employment: These include those persons who were employed at the time of leaving the place of origin but went to the place of destination in search of better employment, in terms of emoluments, job satisfaction, etc.

To take up employment/better employment: In these cases, the persons went to the destination not for searching of employment but were offered jobs or were offered better jobs than they were having at the time of leaving the place of origin.

Proximity to place of work: The persons who had moved in order to be nearer to their places of work.

Studies: Students and others who leaved their usual place of residence for studies are classified under this category.

Appendix 3

Capital and revenue expenditure (Rs.) in the Sundarban, 2009–2010 to 2018–2019

Head of account	**2009–2010**	**2010–2011**	**2011–2012**	**2012–2013**	**2013–2014**	**2014–2015**	**2015–2016**	**2016–2017**	**2017–2018 (revised estimate)**	**2018–2019 (budget estimate)**
Area development	168,775,204	–	–	–	–	–	–	–	–	–
Special component plan for scheduled castes	452,890,014	186,249,474	189,219,594	380,457,417	287,587,542	63,894,293	96,076,346	659,853,056	631,752,000	1,273,275,000
Tribal areas sub-plan	72,474,763	28,814,026	25,148,706	39,095,261	35,606,916	67,638,403	14,398,009	65,493,513	1,701,692,000	2,626,725,000
Other expenditures	688,659,651	535,861,612	347,755,130	401,307,216	240,279,863	289,820,730	360,551,126	612,169,538	50,000,000	–
Gross capital expenditure	**1,382,799,632**	**750,925,112**	**562,123,430**	**820,859,894**	**563,474,321**	**421,353,426**	**471,025,481**	**1,337,516,107**	**2,383,444,000**	**3,900,000,000**
Area development	338,202,420	374,331,112	703,939,833	930,287,143	871,882,784	853,464,698	1,073,948,648	1,474,270,395	434,648,000	597,875,000
Special component plan for scheduled castes	56,550,739	117,919,285	246,613,058	366,243,889	653,451,350	900,293,400	1,099,856,629	773,895,914	190,425,000	308,400,000
Tribal areas sub-plan	3,364,391	12,729,634	33,313,479	50,637,276	89,011,951	117,027,587	216,347,654	26,151,547	57,525,000	157,100,000
Other expenditures	–	–	–	–	–	–	–	–	–	–
General expenditure	-11,227,339	-645,532	-184,097	–	-277,550	–	–	–	–	–
Gross revenue expenditure	**386,890,211**	**504,334,499**	**983,682,273**	**1,347,168,308**	**1,614,068,535**	**1,870,785,685**	**2,390,152,931**	**2,274,317,856**	**682,598,000**	**1,063,375,000**

Source: DoSA, 2011-2018

MIX
Papier aus verantwortungsvollen Quellen
Paper from responsible sources
FSC® C105338

If you have any concerns about our products,
you can contact us on
ProductSafety@springernature.com

In case Publisher is established outside the EU,
the EU authorized representative is:
Springer Nature Customer Service Center GmbH
Europaplatz 3, 69115 Heidelberg, Germany

Printed by Libri Plureos GmbH
in Hamburg, Germany